SMART INSTANT PAYMENTS

TRANSFORMING COMMERCE AND BANKING IN
THE NEW ERA OF US INSTANT PAYMENTS

BY STEVEN WASSERMAN, ELSPETH BLOODGOOD,
ATTILA CSUTAK, PETER DAVEY, TRAVIS DULANEY, REED
LUHTANEN, KEVIN OLSEN, SEAN RODRIGUEZ, CONNIE
THEIEN, MICHAEL E. YOUNG

PUBLISHED BY
PHOTON COMMERCE

Contents

Contents

Foreword

Sean Rodriquez - Retired Executive Vice President - Faster Payments Strategy Leader, Federal Reserve System

In my 35 plus years at the Federal Reserve, I participated in several key US payment system enhancements that were aimed at improving the speed of funds exchange between payers and payees. These efforts included: the maturation of the Automated Clearing House that now includes same-day clearing capabilities; implementation of the Expedited Funds Availability Act (EFFA), which sped up the banking system's exception processing for check payments; and the implementation of the Check 21 Act, which enabled the exchange of digitized check images eliminating reliance on physical transportation of checks.

The capstone of my career was carrying out the industry-developed desired outcome for, once again, improving the speed of the US payment system. I had the honor and privilege to lead the Faster Payments Task Force, which was a broad and inclusive group of over 300 payment industry stakeholders with representatives from organizations across the US payment ecosystem, including financial institutions, non-bank payment providers, regulators, standards bodies, consultants, businesses (merchants and corporates), and consumer groups. The Federal Reserve assembled the task force in May 2015 to collaboratively identify and evaluate alternative approaches to implementing safe, ubiquitous, faster payments capabilities in the United States.

Faster, or real-time payments was a relatively new concept in the US in the 2015 timeframe, but one which we had observed and learned from other instant payment platforms that had already emerged in other parts of the world. Some US-based non-bank payment platforms had already started to emerge where funds could be immediately used by the payees, but for those transacting through bank accounts, they took up to 2-3 business days.

The pressure was on to develop a more holistic approach for the country that would serve needs well into the future. The Faster Payments Task Force focused its efforts on identifying goals and attributes of effective faster payment systems; proposing solutions and assessing their capability to achieve those goals; and championing the US payment industry to take steps toward implementation and adoption of faster payments capabilities. Prior to my retiring from the Fed in early 2019, we had set the stage for a new realm of payments; instant payments that would immediately provide available funds and simultaneously settle between the financial institutions of payor and payee of transactions.

I was proud to be a part of our nation's collective efforts to design and implement faster payment capabilities for our country. Now, I am thrilled to endorse the collaborative effort put forth to develop this book that documents the faster/instant payments history and that explains what payments practitioners should know about:

- The history of faster/instant/real-time payments in the US

- What instant payments are

- How they work and differ from other payments

- Who are the stakeholders that enable it and for who will this be used by

- Why are they important and what they can be used for (.i.e. the use cases)

Through the network of industry stakeholders that I interacted with I was introduced to Michael Young, Founder and CEO of Photon Commerce who was sponsoring this book by bringing together an elite group of thought leaders and subject matter experts involved in various industry collaborations.

- Elspeth Bloodgood, AAP, NCP - Technical Product Manager Senior Advisory, Jack Henry Payment Solutions

- Attila Csutak, PMC - Faster Payments, Vice President, City National Bank

- Peter Davey - Executive Leader | Innovator | Strategist | Influencer | Payments & Banking | Payments Jedi, Alloy Labs

- Travis Dulaney - Founder | Payments | Risk Management | Fintech | Advisor | Venture Builder | Real Estate | Angel Investor | Speaker, CEO and Founder, BalancedTrust

- Reed Luhtanen - Executive Director and CEO, U.S. Faster Payments Council

- Kevin Olsen, AAP, NCP, APRP, CHPC, SVP - Innovation & Strategy | Payment$ Professor | Executive Coach | Inspirational Life Speaker, Pidgin

- Connie Theien - Senior Vice President, Payments Industry Relations, Federal Reserve System

- Steven Wasserman - Technology Architect, Photon Commerce | CEO/Founder/Architect, Vments | Board Advisory Group, US Faster Payments Council | Payments Wizard

Michael's leadership in assembling this group of experts to deliver this book for the benefit of payment practitioners in the US and beyond has been exemplary. Further, Michael's vision and work at Photon Commerce is enabling a world of instant payments leveraging the latest of technologies to provide optimal user experiences and effortless automated payment processes. These are many reasons why you should read this book and learn from the collective experience, expertise, and guidance of these esteemed professionals and industry leaders.

Are you ready for faster/instant/real-time payments? Don't miss the opportunity to be a part of this transformational and exciting time that will be marked in the history of payments.

Preface

We at Photon Commerce have been collaborating and establishing relationships with others across the payment industry through member organizations such as the US Faster Payments Council, the Federal Reserve's FedImprovements initiated Business Payments Coalition, and other payment industry membership organizations.

Through these organizations, we have selected a group of key thought leaders and subject matter experts that we've brought together to assemble this collection of topics of interest, education, and necessary information for those that are looking to implement and/or use instant payments.

We are proud to have sponsored this effort to bring you this book.

About Photon Commerce

Photon Commerce has been helping various stakeholders in the payments industry to facilitate and accelerate the adoption of instant payments by leveraging its AI powered document data capture capabilities off of documents such as invoices, purchase receipts, purchase orders, shipping and receiving documents, contracts,, checks, payment remittance detail lists, and other documents related to the origination or reconciliation of instant payments.

Who was this book created for

This book is tailored for a diverse audience, including payment practitioners, financial institution (FI) executives keen on implementing instant payments, FI product and business analysts in the process of designing implementation roadmaps, fintech companies, corporates, and other interested applicable parties.

Why we wrote it

The reason behind creating this insightful piece stems from the evident lack of education and awareness regarding instant payments. We aim to bridge this knowledge gap, igniting the drive for implementation and adoption while offering valuable planning considerations and guidelines.

What to expect and benefit from reading this

Delve into an understanding of what instant payments truly entail and why they are crucial in today's financial landscape. Explore the social, economic, and philosophical implications of this revolutionary payment system, tracing its history both on an international scale and within the US. Understand the pivotal role of the ISO 20022 standard, its features, functionalities, and overarching benefits. Dive into a multitude of use cases, examples, and implementation considerations. Additionally, gain insights into risk, fraud, and compliance considerations.

We then discuss potential rocket boosters of instant payment adoption and enhancement, encompassing AI, QR code technology, Open Banking APIs, Embedded Payments, and directories

We close by taking a look into the future of instant payments.

Why you should read this

If you're curious about what may seem like mere hype to many, this book is your gateway to understanding the profound implications and potential of instant payments. As a transformative force comparable to the origins of money and payment methods like checks and cards, instant payments stand at an early but immensely impactful stage. By reading this book, you'll equip yourself with the necessary knowledge to either kickstart your instant payment journey or re-evaluate ongoing implementations. Don't risk being left in the rearview mirror, potentially losing customers, deposits, and revenues to other institutions or services.

Join us on this informative journey to embrace the future of payments - read, learn, and take action to remain at the forefront of this dynamic evolution in finance.

Introduction

"The generation of NOW and ALWAYS" is where everyone expects everything to be done instantly and available 24x7x365".

There were days when banks and many retail businesses were only open on weekdays excluding holidays that fell out during the week and we were content and patient with that for the most part as we understood that this was the norm at the time.

That started to change as many banks and retail businesses started.to be open 6 or 7 days a week. On holidays such as "black friday" after Thanksgiving, business opened for shopping in the middle of the night. This expanded to staying open during Thanksgiving all night to the next day and in some cases straight through the weekend and even some that went straight through the end of the end of year holiday season. Then the "black friday" and hours started earlier in the fall and the after Christmas deals found their way into January and started to blend with the President's Day sales and extended hours.

It used to be rare to see any retail businesses open 24x7, and fewer 24x7x365, but there are many examples now which we can find open all the time such as some gas stations, convenience stores, pharmacies, groceries, fast food restaurants, and others.

Today we can also find many banks open on Saturdays. With the introduction of ATMs, limited banking services became available 24x7x365.

The always available internet enabled e-Commerce at any time where we could place an order in minutes and pay for it instantly by card accounts which were debited instantly as well, but the banks settled these between them behind and then credited the merchants up to 3 business days later.

Through advancements by companies like Amazon, UPS, Fedex, and others, the shipment of e-Commerce and delivery of packages in general started to shrink expectations from days or weeks down to as soon as same day receipt.

The current generation that did not know of anything prior to these always open and instant everything has established a new norm, that we also now expect for payments. For those who have in their payments life, only known of digital card and non-bank payment services (e.g. PayPal and Venmo) likely do not even know what checks are or don't understand why anyone would use them where it could take several days for checks to clear and be fully available.

When the checks are sent via the mail, it takes that much longer before they even get deposited. Some businesses may say that checks give them time before the funds come out of their account, but the uncertainty of how long this might take and the effort it takes to reconcile checks and calculate cash flow taking uncleared checks into account may be fooling themselves.

During Covid, when offices closed and many were forced to work at home, and many feared the spread of Covid by handling the paper, the printing and deposit processing of checks by businesses accelerated the switch to electronic payment methods.

The switch over from cash to electronic payments also accelerated during Covid. It can be viewed that cash is a form of instant transfer of value especially when it can be immediately re-used. However, if the cash has to be deposited into bank accounts before it can be used, then cash falls short of the benefits of instant payments where the

9

payee can use the funds immediately. Possession of cash also comes with other issues that do not exist with electronic payments such as losing it, having it stolen, not able to accrue interest, only able to use for in person transactions, and not always accepted everywhere.

Some uses of checks and cash switched to instant payments during Covid, but the instant payments implementations had not matured enough and alternatives such as ACH and card payments were more of the beneficiaries of forced switches from checks to electronic payments during this period. Even with this acceleration, B2B payments still have a long way to go to wean off of checks and as the saying goes "cash if king" it will likely still linger on a continued decline, but not for some of the reasons that instant payment adoption of instant payments will accelerate.

Instant payment network implementations can't come fast enough for many financial institutions as they begin to compete for the payment landscape with non-bank financial services (e.g. PayPal, Venmo, and Crypto payment platforms) that on the surface appear to be instant and always operating to move funds between payers and payees as they provide "closed-loop" platforms and ecosystems where banks have only been able to do the equivalent with payers and payees that are their customers on both ends. These "on-us" and "closed-loop" transactions settle immediately between payer and payee and do not require settling funds with other financial institutions or networks other than when the non-bank "closed-loop" users load funds into the loop or unload them back out. Loading and unloading funds is called on-ramps and off-ramps and interestingly enough these have become one of the top use cases for instant payments where users of these non-bank closed-loop platforms could instantly and automatically transfer funds with their bank accounts and these non-bank platforms.

The demand is there and if your financial institution and payment service customers are not asking for instant payments, it could be because they are already using something else or these services have not asked them the right questions or offered them instant payment use cases that they would better understand the value of, such as instant payroll, instant merchant settlement, and many other use cases described in this book.

The time to act is now and always be re-evaluating and improving to first catch up and then expand in the world of the NOW and ALWAYS of instant payments.

There are a number of instant payment specific terms and stakeholders that the book mentions and which you should be familiar with. Let's start out by first defining faster payment and then further distinguish instant payments.

Note: The US Faster Payments Council (FPC) published a Glossary of Terms for the industry. Many of these terms were collaborated together with a team of subject matter experts from the NACHA Faster Payment Professional Certification work group. Applicable references to the glossary work group as well as other sources are accordingly footnote referenced.

- **Faster Payments:** Electronic payment services that provide funds to the Payee within seconds or up to a few hours. Faster payments include instant/immediate/real-time, push-to-card, and same day ACH. [1]

- **Instant Payments:** An electronic payment solution available 24/7/365, resulting in the immediate interbank clearing of the transaction and crediting of the Payee's account with confirmation to the Payer within seconds of payment initiation.[1]

Here are a few other terms and stakeholders you should become familiar with...

- **Credit Push Payment:** A payment initiated when a Payer sends a payment order with instructions to transfer funds to the Payee.[1] Instant payments in the US only implemented credit push payments versus debit pull type payments that occur with debit cards and ACH transactions.

- **Credit Transfer:** A payment transaction that increases an account balance.[1] The instant payment use of ISO 20022 credit transfer messages are the credit push payment money movements in the instant payment networks use of ISO 20022. Even a return of funds is a credit transfer back to the original payer's account.

- **Debit Pull Payment:** A payment made after prior authorization by the Payer. The Payee sends the payment instruction to the Payee's account to draw on funds from the Payer.[1] The instant payment networks in the US do not support any debit pull payments which in some other countries are supported by where they are referred to as Direct Debits.

- **Deferred Net Settlement:** A settlement option that settles on a net basis at the end of a predefined settlement cycle typically at the end of the business day, but sometimes during the business day.[1] The instant payment networks settle using a method called Real-Time Gross Settlement, which is defined below. The card networks and ACH use a deferred net settlement method.

- **FedNow:** The FedNow Service is a new instant payment infrastructure developed by the Federal Reserve that allows financial institutions of every size across the U.S. to provide safe and efficient instant payment services. [2]

- **Fintech:** Any business that uses technology to modify, enhance, or automate financial services for businesses or consumers.[3] Many financial institutions have partnered with and in many cases have acquired Fintechs to air in the implementation and enhancements to their payment offerings, including instant payments.

- **Good Funds:** The funds that are unconditionally available to the owner of the receiving account and usable immediately by the owner of the account.[4] Instant payments follow a good funds model.

- **Interbank Settlement:** The final settlement between banks done within seconds of an instant/immediate/real-time payment or through a deferred settlement scheme where funds are available to the receiver either before (i.e., push-tocard) or after the bank-to-bank settlement (Same Day ACH).[1]

- **Interoperability:** A process that enables solutions to transmit and receive payment instructions across various payment systems or platforms. It requires the use of common applied technical standards, coordinated digital identities, alias directories, and broad access to settlement mechanisms compatible between systems.[1]

- **Irrevocable Payment:** A payment that is final and typically has no recourse for correction or reversal.[1] Instant payment transactions are irrevocable, though through the US network specific rules, there are some exceptions, such as in the case of error or fraud where a request for return of funds can be used versus where in ACH and card networks, transactions can pull the funds back.

- **ISO 20022:** An international standard for exchanging electronic messages between financial institutions. First introduced in 2004, ISO 20022 was created to give the financial industry a common platform for developing messages using a modeling methodology, a central dictionary, and a set of XML and ASN.[1] design rules.[5] The instant payment networks use this standard for its rich data transaction messages.

- **ISO 8583:** An international standard for financial transaction card originated interchange messaging. It is the International Organization for Standardization standard for systems that exchange electronic transactions initiated by cardholders using payment cards.[1]

- **Nacha (National Automated Clearing House Association):** The entity which manages the development, administration, and governance of the ACH Network.[1] Nacha has collaborated with the US Faster Payment Council to create a Faster Payments Professional Certification program as well as a Accredited Faster Payments Professional (AFPP) test.

- **OFAC (Office of Foreign Assets Control):** An organization that administers and enforces economic and trade sanctions based on US foreign policy and national security goals against targeted foreign countries, terrorists, international narcotics traffickers, and those engaged in activities related to the proliferation of weapons of mass destruction. OFAC acts under Presidential wartime and national emergency powers as well as authority granted by specific legislation to impose controls on transactions and freeze foreign assets under US jurisdiction.[6]

- **Participant:** Someone who has legal access to a payment solution, payment network, or payment service.[1] For the instant payment networks, these are the financial institutions and any third party service they used to connect to these networks.

- **Payment Initiation:** A process that is triggered when either the Payer or Payee in a payment transaction, or a third party, sends an instruction to another entity that ultimately leads to a payment. The initiation ends at the point when the Payer authorizes a payment order, or in the case of pre-authorization, when the provider confirms that pre-authorization exists for a given payment.[7] Instant payments initiation is where a payer initiates a credit push which may or may not have been in conjunction the the payee initiating a request for payment.

- **Push-to-Card:** A process whereby money is sent to a debit or prepaid card belonging to an individual or business.[1] Even though push-to-card transactions do provide near-real time availability of funds to the payee, the interbank settlement behind the scenes uses the deferred network settlement method described above.

- **Real-Time Gross Settlement:** The settlement of payments, transfer instructions, or other obligations individually on a transaction-by-transaction basis.[1] This is the settlement method used by instant payments.

- **Real-Time Payments:** A real-time payment includes the transmission of the payment message and the availability of final funds to the Payee, occurring in real time or near real time, and on as near to 24-hour and seven (7) days basis as possible.[1] Instant payment networks are also referred to as real-time and immediate payment systems.

- **Remittance Data:** Information the Payer provides to the Payee about the reason or details of a payment.[1] Instant payment use of ISO 20022 enables including remittance data with the payment, sent separately as a standalone message, and is included in some of the inquiry and reporting ISO 20022 messages.

- **Request for Payment (RfP):** A standardized network message that Payees can leverage to send electronic requests to Payers through the network.[1] Instant payment networks include use of this ISO 20022 message type and is a topic discussed through many of the chapters of this book.

- **RTP:** This is a trademark for the real-time payments network owned and operated by The Clearing House.[1] The Clearing House (TCH) is a not-for-profit organization owned and operated by many of the US largest banks. Their RTP® network is one of the two instant payment networks in the US.

- **Same Day ACH:** An ACH payment following the Nacha rules, that is sent, received, and settled the same business day.[1]

15

- **Ubiquitous:** In reference to payments, it is a system in which any Payer can make a payment to any Payee.[1] This is the ultimate goal of interoperability of instant payment networks where any payer can transact an instant payment with any payee. As of the date of this book, we have yet to have ubiquity in either of the US instant payment networks nor are they interoperable between financial institution accounts that are on just one of the two US instant payment networks. As implementations of these networks mature, and all US financial institutions are on at least one of the two networks where many will also be on both, the goal of ubiquity is within sight.

- **US Faster Payments Council (FPC):** The Faster Payments Council (FPC) is an industry-led membership organization whose vision is a world-class payment system where Americans can safely and securely pay anyone, anywhere, at any time and with near-immediate funds availability. By design, the FPC encourages a diverse range of perspectives and is open to all stakeholders in the U.S. payment system. Guided by principles of fairness, inclusiveness, flexibility and transparency, the FPC uses collaborative, problem-solving approaches to resolve the issues that are inhibiting broad faster payments adoption in this country.[8] This not for profit organization was stood up following an industry collaboration organized by the Federal Reserve in an effort called the Faster Payments Task Force.

- **Zelle®:** is an easy way to send money directly between almost any U.S. bank accounts typically within minutes. With just an email address or U.S. mobile phone number, you can quickly, safely and easily send and receive money with more people you know and trust, regardless of where they bank.[9].The US Instant payment rails moves funds between bank accounts in seconds versus minutes, but they currently require bank routing and account numbers until applicable directories support the use of aliases like those used by Zelle, PayPal, and Venmo. Interbank settlement of Zelle transactions is not done instantly as it is for instant payments. Zelle is interbank settled using ACH and card network settlement methods, though it has started adding instant payment rail settlement as an option.

Now that you have some terminology mastered, you have a base for better appreciating and understanding where these terms and stakeholders are referenced through the book.

We hope you're ready to learn and then act on implementing or expanding existing implementations of instant payments in the US. By reading this book; you should come away with the following:

- Knowledge about various aspects of what instant payment are, how they originated, who and how they benefit financial institutions, Fintechs, payment service providers, and the consumer and business end users

- Social and philosophical implications of instant payments

- Economic effects and considerations of implementing and/or adopting instant payments

- Use case benefits and corresponding implementation considerations

- Implementation considerations including strategies, use cases, risks, fraud, compliance

Chapter 1 - Societal, Economic, and Philosophical Implications of Instant Payments

By Reed Luhtanen

"…a foundation for the future—a modern payment infrastructure that allows innovation and competition to flourish…" Governor Lael Brainerd, Federal Reserve Board of Governors[1]

The development and adoption of instant payments technology will fundamentally change how our economy operates in the future. I guess this might come across as hyperbolic, or sound like a bit of industry grandstanding, but I really believe it's true. And I'm going to tell you why.

The very functioning of the modern economy depends on the electronic movement of money from payers to payees. Without that ability the entire world economy would stop functioning. Further, as technology has evolved in all aspects of our lives, the functioning of digital payments has continued to grow in importance in lock step. But it's important to know that this is not a binary "do we have electronic payments?" or "do we not have electronic payments?" No. There is a continuum of payments. And just as the lack of a functioning system of electronic payments would torpedo the global economy, so too can the advent of wholly new, 21st century payments technology catalyze a new era of economic development, empowerment, and inclusion.

A powerful new payment type can unlock new uses for payments, making the development of entirely new business models possible. I will discuss this in more depth later in this chapter, but think about the ways previous innovations in payments opened up new channels of commerce and actually empowered people to be more mobile, more social, more industrious. Every advance in the transfer of value (including the jump from trading goods and services to the development and use of money to represent value) has unlocked wide-ranging and unpredictable leaps forward in the ways people are able to structure and conduct not just business relationships but how they live their lives.

This next leap forward will have the same profound, and unpredictable, effects on our economy and our society more broadly. And this will come not only from the enablement of new business models (in much the same way the combination of the internet and payment cards made business models such as Amazon.com and Netflix possible), but in the empowerment and inclusion of previously under-represented populations. This will be discussed at length later in this chapter, but the prospect of ensuring broad inclusion of previously underserved groups is a compelling reason to get behind and champion this leap forward in payments.

But before we get into that, we need to zoom way out. Those of us in the payments industry have a tendency to jump right into the weeds. The clearing and settling. The recourse rules. The message specifications. That is all important! But equally important is understanding exactly where payments fit into the broader context of our daily lives, and the daily lives of people who don't have any reason to know what ISO 20022 is, or the difference between RTP® and FedNow®.

Payments, at their core, are about relationships. A lot of infrastructure – plumbing, electricity, roads – they exist to enable both communal and individual pursuits. When I get in my car to drive somewhere, it might be that I'm going to meet up with someone for a beer. It might be that I'm going to a store to buy something (probably beer). Those, of course, are communal activities – multiple people meeting socially or engaging in commerce. On the other hand, maybe I'm going to the lake to be by myself. Maybe I'm just going for a drive, and I don't have a destination in mind.

Payments are different though, because a payment only happens when two parties have some other relationship that results in an obligation flowing from one of the parties to the other party. And contrary to popular culture representations of people lamenting their bills or trying to get their friends to pick up the bar tab, the reality is that in most situations both the person paying, and the person being paid both want the payment to happen. They both benefit from it.

So, it's important to understand that payments are, by their very nature, a communal activity with, at a minimum, two parties involved. Now, we don't think of ourselves as "parties to a payments transaction." We think of ourselves as friends who owe or are owed money. We are customers who want to buy beer, or we're merchants who want to sell beer. We are manufacturers who make the beer (yes, I know there are goods other than beer, but none of them are as good as beer), distributors who want to take the beer from the manufacturer to the retailer, thirsty consumers returning empty beer cans and getting their deposits back. Whatever persona we are in the various payments we're involved in, all payments are communal.

And, of course, if you want to get deep into the language for a minute (I would guess everyone who picked up a book about faster payments really wants to learn about etymology), the Latin root of communal, communis, means "common, of the community." Not really a surprise, but communis is also the root of communication. Payments aren't just a way of transferring value from one party to another. Payments are communication. And the advent of instant payments sets up, for the very first time, the opportunity to fully communicate and transfer value associated with that communication, in real-time.

I'm not claiming to be the first person to have this thought. Far from it. But it's critical to understanding the potential impact of instant payments on our social structures. In her book, *New Money: How Payment Became Social Media*, Prof. Lana Swartz from the University of Virginia, notes that "one of the classic Aristotelian functions of money is to serve as a 'medium of exchange,'...and Immanual Kant found money to be 'the greatest and most useable for all means of human communication through things."[2] For her part, Swartz helps put this concept into our modern context:

Transactions are embedded in and reflective of social, cultural, and relational meanings. These meanings shape and are shaped by the communication and media technologies—whether paper or electronic—that perform the transactions.[3]

You see, how we transact is one way that we communicate to the world, and the various forms of payment are signals that we all use to infer things about the social standing of others. Seeing a person pay with a check, cash, debit card, credit card, or mobile phone will immediately cause us to conclude things about that person that aren't directly related to that choice. In fact, I would wager that everyone reading this has made a payment choice based not on a purely rational assessment of the financial implications, but on what that choice communicates to someone else. But payments also necessarily communicate other important messages between the payer and payee as well. In some cases, very detailed messages.

Payments, then, are fundamentally a medium of communication – messages (in the broadest sense possible) sent and received. But these messages themselves are inherently valuable, not only because of the information provided but because money is moving along or linked with that information.

I hope it's clear by now that it's going to get a little weird in this chapter. Because this chapter is about what this all means. It's about how advancements in communication technology and media influence our lives and our communities in profound ways, so that's where we will start. But then we will look into how advancements in payments – from paper, to batch electronic, and now to instant payments – have brought about incredible and unpredictable changes in how we conduct our lives and the business models that are even possible. After that, we will take a look at how modern payments will alter the course of payments as a mechanism for social divisions, how this foundational technology will bring more people into the modern economy and could serve to eliminate, or at least reduce, class barriers that are often reinforced by payments.

Section 1 – A Brief History of Communications

"The medium is the message"

- Marshall McLuhan, Understanding the Media: The Extensions of Man

As I begin this section, I want to dispel any notion that what follows here will be an argument that the evolution we are experiencing is something wholly new. That is simply not the case. Instead, we have to think of this as an evolution, not a new technology built from scratch. For those of us who are interested in thinking forward to if, how, and when this evolutionary technology will impact various aspects of society, this is helpful. The same has been true of other advances in communication technology. To borrow from Marshall T. Poe:

First we spoke. Then we wrote. Then we printed. Then we listened to the radio and watched TV. And now we surf the internet. Each of these media was different from the others, but all of them were of a piece - tools that we used to send, receive, store, and retrieve messages. The Internet, it seemed to me, was not so much brand new as a variation on an ancient theme.[4]

Likewise -- and while we will explore this in Section 2 of this chapter I want you to have this in mind now – payments systems have gone through this same evolution, and in many ways advancements in payments happened in concert with advancements in other forms of communication, but not to the extent you might assume. With that in mind, let's dig into the history of communication. I know this is why you bought this book, after all.

As you no doubt anticipated, this section of the chapter will begin with a brief discussion of Homer.[5] Up until the 1930s, it was believed that Homer was in fact a great writer, who personally penned the timeless classics *The Iliad* and *The Odyssey*. That was when two scholars, Milman Parry and Albert Lord turned the world of Classical Scholarship on its head by demonstrating that not only was it possible, it was likely that Homer was an oral tradition, not a single person who was a great writer. Parry and Lord believed this because the structure and lyricality of the work strongly suggested that it was sung, not simply written.

The interesting thing about this revelation, at least for our purposes, is not the fact that Homer wasn't some guy who wrote 1,000 page poems. The really interesting thing is how Parry and Lord went about demonstrating the likelihood of their theory. You see, the big question wasn't about finding evidence that a person did or did not exist. That would be impossible, given the lack of those sorts of records from that time period. The real question was whether something as cohesive, lengthy, and beautiful could possibly have been compiled and then passed down orally before eventually being written down. To prove that, they looked not to the past, but to the (then, remember this was almost 100 years ago) present, traveling to the Balkans where they found and recorded singers who sang such long heroic tales.

That revelation proved immensely influential on another scholar, Harold Allen Innis. Innis took that finding from Parry and Lord, and extended it out. Ultimately, what Parry and Lord had demonstrated was that communication technologies – in the case of the Homeric works, speech – shape the content. Innis took this a step further in arguing that the effects are broader, that communication technologies not only shape content – they shape the very society in which they are deployed.[6]

So if the medium of communication shapes the content, and even shapes the very society it's operating in, we should think about how we will assess that communication medium based on various attributes. Again, Poe helps us by breaking down the significant media attributes as follows:

- Accessibility: the availability of the medium itself.

- Privacy: the covertness with which data can be transmitted in a medium.

- Fidelity: the faithfulness with which data can be transmitted in a medium.

- Volume: the quantity in which data can be transmitted in a medium.

- Velocity: the speed with which data can be transmitted in a medium.

- Range: the distance over which data can be transmitted in a medium.

- Persistence: the duration over which data can be preserved in a medium.

- Searchability: the efficiency with which data can be found in a medium.[7]

As you read that, you might have been thinking about the fact that just about any communication method can be made more or less accessible, or to have higher or lower fidelity. And that's probably correct, at least to a point (for example, describing a physical object with words will only get you so far in terms of fidelity). So when considering these attributes, we should do so through the lens of the costs associated with increasing each attribute. So, thinking about range, for example. Both a physical letter and an email have theoretically the same range. Both can be delivered anywhere in the world. But in the case of a physical letter, it might be very expensive to deliver it from New York to a remote location on another continent. In the case of email, the cost of delivering it next door or across the world is basically the same

Full disclosure, having read Poe in preparation for writing this chapter, I think it's safe to say that I had not previously fully considered what it means to say that payments are a communication medium, and had certainly not assessed various payment methods using these attributes, but if you haven't already figured this out – that is where we are heading in this chapter. For now, keep these attributes in mind as we continue looking at the evolution of communication technology and media over the course of human history.

Writing

It might seem strange to think of writing as an invention, or as technology, since it seems so obvious and automatic to us. But before I go back in time, give this some thought. It takes a relatively long time to learn to read. Unlike walking and talking, which typically begin around 18 to 24 months into life for people born in the United States, reading typically is not mastered until a child is 7 or 8 years old, when they transition from "learning to read" to "reading to learn" around second grade. In fact, humans existed for about 175,000 years before writing was invented, which was roughly 5,000 years ago. Said another way, for 97% of human history there was no writing.

Writing wasn't invented until that time period, around 3,000 BC, not because humans lacked the technology to make markings. Humans had long been carving and painting, both of which demonstrate the sorts of technology that enable writing. Instead, it was around this time that humans transitioned from migratory societies to living in villages. It did not take long before villages led to government, which led to tax collection. It was this practice – the collecting of taxes and thus the need to keep accurate accounts of who had paid and how much – that pulled writing into existence. [8]

Thus, the fact that you're reading this book about payments innovation right now, the fact that you can read, that the written word exists at all, is because of payments. That's not exactly important here, but it is fun to know.

Printing

Everyone knows that printing was invented by Johann Gutenberg, who famously invented the printing press and immediately began printing Gutenberg Bibles in 1455. Except that isn't really how it went down. The technology needed to print had existed for some time, and indeed, presses had been in use in various parts of the world prior to Gutenberg's time. So what gives? Why did we all learn that Gutenberg was such a big deal? The fact is that the world wasn't previously ready for the amount of written material that a movable type printing press would unleash upon it.

That began to change in Europe in the fifteenth century, as three distinct groups – capitalists, bureaucrats, and pastors – became dissatisfied with handwritten manuscripts, and each began demanding (and, in turn, promoting) the use of print.

As European society became increasingly organized and predictable in the fifteenth century, with political power consolidating into city-states and then full-blown nation-states, the development and spread of commerce via entrepreneurship emerged. These entrepreneurs didn't just buy and sell goods, though, they developed innovative financial institutions: the joint stock company, commercial bank, insurance agencies, and more. These businesses all required standard contracts, forms, and other printed materials that required mass production (and if you look at some of the forms and processes in 21st century financial institutions you will wonder if they have changed at all since the 15th century).[9]

At this same time, you had the rise of a bureaucratic state, which was concerned largely with the development of military power to safeguard the respective sovereignty of the various nation-states that were emerging at that time. There was also the emergence of religious dogma, principally championed by Martin Luther, that involved worshippers in their own learning, which further necessitated access to the written word.

So what were the implications of the widespread adoption of this media? While we covered a few discernable use cases that pulled printing the written word into prevalence, those specific things aren't the reason printing changed the world. Rather, they were the gateway – the thing that got that technology over the hump. Once it was out there, the effects literally changed the way people lived.

Largely due to the tremendous increase in accessibility to written works, mass literacy rapidly emerged throughout Europe and then spread to the rest of the world. By the 18th century, more than half of the children in Northwestern Europe had the opportunity to attend primary and secondary school. Around 1500, about 1 percent of Western Europeans attended university. A century later that number had increased to 2.4 percent in England and 2.8 percent in other parts of Western Europe. By 1800 more than 50 percent of adult males in Northwestern Europe could write their own name, and more than that could read simple passages.[10]

It's probably not a surprise to learn that the availability of the world's first mass media, the printed word, led to dramatic increases in literacy and more people (at least more men) attending university. But that's not the end of the story. The ability to print and then transport written words opened up channels of political dialogue that led to the Protestant Reformation and the American Revolution (not to mention the English and French Revolutions). The printed word, and the reasoned debates that it enabled gave rise to the limited monarchy and out of that construct western liberal democracy was birthed.[11] New forms of communication change the world in profound, unpredictable ways. And we are just scratching the surface.

So a quick check of the timeline of human communication:

- First 175,000 Years: Speech only
- 5,000 Years Ago: Written Word Emerges
- 500 Years Ago: Printed Word Emerges[12]

Computing and the Internet[13]

While the internet is clearly the most advanced of the three forms of written communication that I'm covering in this chapter, because most of us lived through its widespread introduction and expansion into all facets of our daily personal and professional lives it also likely requires the least in terms of background for you, our reader. That said, I think it's interesting (and more importantly, I think you might think it's interesting) to look into a couple of the three demands that pulled the internet into such widespread use: information capitalism, the surveillance state, and cultural privatism.[14] The first two are most relevant to our focus here, so I will only be discussing those below.

Taking them in that order, then, it's worth noting that computing machines and the idea of connecting them had existed for quite some time before the tech boom of the 1990s. In fact, calculating "business machines" had been used in business since at least the 1890s. Generally, though, those machines were too expensive for most businesses to use. Then, with the advent of time-sharing through dialup connections in the 1960s, more businesses began to see the value in computing. Finally, though, it was the introduction of the personal computer in the 1970s that made mass production of computers happen. So, while businesses were not the primary driver of demand for the internet, "once they saw it for what it was – both a new place and a new way to conduct business – they very quickly exploited the opportunity it provided, thereby hastening its expansion."

The second major pull for the internet, according to Poe, came from the surveillance state. There were a couple different reasons for this. First, the modern welfare state required tremendous amounts of data about the citizens to administer various social welfare programs. If you are going to send Social Security payments to everyone over the age of 66, you need to know a lot of information about everyone in the country, for example. Additionally, governments recognized the internet as extremely useful for national defense purposes. Again, per Poe, "[t]he U.S. Defense Department funded virtually all of the research that led to the production of the first American computers, the first American computer networks, and the immediate forerunner of the Internet, ARPANET."[15]

The effects of the internet are difficult to overstate. It's transformed every facet of our lives. We consume content, whether newspapers, magazines, television shows, films, family photos, you name it, all via the internet. We communicate via email and text messages and Zoom. And it has completely transformed modern commerce, notably retail commerce, in ways that were certainly not predictable when the Department of Defense was funding its development. As I sit here writing this on a computer I ordered from Apple.com, looking at a monitor I ordered from Walmart.com, sitting on a chair I ordered from Target.com, at a desk I ordered from PotteryBarn.com, on a rug I ordered from ruggable.com listening to music streaming from Spotify,[16] it's conceivable that the confluence of the internet as communication media and the internet as a commercial platform is complete.

Or is it? Even now, the dramatic expansion of connected devices continues. Our washing machine pops up an alert on our phones and television screen when the cycle is complete. Our ceiling fans and lights are controlled by an app on our phones or by voice command to Alexa or Siri. Our doorbell alerts us not only when someone presses the button, but when someone, or something, walks nearby (the source of more than one video of deer eating our landscaping). Increasingly these devices, too, are connecting into channels of commerce, but imagine the innovations that must be on the horizon.

None of this would be possible, of course, without the combination of new technologies – Wi-fi, high-quality cameras, sensors, GPS, etc. – and payments. Without modern payments, the internet never could have emerged as the powerhouse of commerce that it has become. And in many ways we are still scratching the surface of the ways the internet (and even more advanced tech) and 21st century payments can usher in even more innovation. That's what we're going to get into in Sections 2 and 3 below, so buckle up.

Section 2 – Payments Innovations Transform the World

The Poe text I used to help structure our thoughts on communications technologies provides a useful framework for looking at the key attributes of payments technology. And while it's not my intent to bore you with the blow by blow for every payment type the world has known, I think it will be instructive (and fun) to look at the ways payments innovations have transformed the world, starting with the development of money as a concept.

Just as it seems strange to think of writing as an invention or innovation in communication technology, it might also seem odd to think of money as such. But the fact is that the first group to mint an official coin as its currency was the Lydians in 600 BCE.[17] This was about 2,600 years ago, meaning that for the first 170,000+ years of human existence there was no money. And this innovation unlocked the potential of Lydia to become an economic powerhouse. How? Money isn't intrinsically valuable. That's the whole point, after all. But consider what money enables that pure trading doesn't.

The most obvious aspect is that it's obviously easier to carry coins than to carry bushels of grain, or animal hides, or barrels of beer. But that's just the obvious piece of the puzzle. The invention of money essentially allowed producers of goods to shift in time when they extracted value from the things they produced. Rather than producing a crop of grain and needing to immediately trade it for other goods or services, lest it spoil, farmers could trade the grain, which will spoil, for coinage that's good forever, and then trade the coins for the goods or services they need at a later time. This wasn't just good for the farmer, though. This encouraged farmers (and indeed all producers of perishable goods) to grow more crops, meaning more people could be fed. Ultimately the development of money went a long way to increasing wealth and reducing poverty and hunger.

At this point, let's pause to assess coins and currency through the lens of the attributes of communication media identified by Poe. These are listed above in Section 1, so feel free to refer back to that discussion for a refresher.

Coins and Currency

Attribute	Definition	Assessment
Accessibility	the availability of the medium itself	High: Anyone could access coins or currency, provided they had goods or services of value to trade for them.
Privacy	the covertness with which data can be transmitted in a medium	High: the exchange of coins or currency could be done in private such that only the parties involved in the transaction would know of it.
Fidelity	the faithfulness with which data can be transmitted in a medium	None: There is nothing that connects a coin or paper note to the underlying transaction, identifies the parties involved, or provides any information about the exchange.
Volume	the quantity in which data can be transmitted in a medium	Low: Physically transporting coins and even paper currency creates a significant burden, particularly as the amounts in question grow in more sophisticated operations and economies.
Velocity	the speed with which data can be transmitted in a medium	Low: The use of coins and currency requires the physical transport of the coins or currency to the location of the exchange. This means the ability to transact is limited by the speed with which that transportation can occur.

Persistence	the duration over which data can be preserved in a medium	Low: Being entirely analog, once the exchange occurs, there is no inherent record of the transaction that is associated with the coins or currency. While it's obviously possible, and common practice, to memorialize a transaction with a receipt of some sort, that receipt is separate from the payment.
Searchability	the efficiency with which data can be found in a medium	None: There is no data to be found in a purely analog system that involves the physical movement of coins or currency from one party to the other.

The arc of advances in payments loosely follows the arc in development of communications technologies. Just as the written word emerged around 5,000 BCE and printing took off around 1450, the first major leap forward in payments after the development of money around 600 BCE occurred around the 9th century. In that era, long-range commerce was growing, and even the transportation of coins was becoming problematic. Traders invented the "sakk," a piece of paper with instructions to the merchant's bank to make a payment from his account. A sakk could be cashed in another city or country, making travel easier and safer from theft.[18] And of course the sakk is the predecessor of modern-day checks.

Checks certainly solved a few of the problems and shortcomings of a purely currency-based economy. Checks eliminated the need to carry large amounts of coins or paper currency for large transactions. They are negotiable, meaning one party can transfer the check to another party, enabling more complex transactions to occur more easily. Later, the advent of clearing houses dramatically enhanced the speed with which financial institutions could process checks, making them more efficient for all involved to use.

Checks also freed people, particularly in the United States, to move about more freely via train and steamboat. In fact, it was largely this emerging aspect of life for Americans, coupled with the growing complexity of business operations, that pulled checks into widespread use. The risk and hassle associated with traveling with large sums of money or gold were eliminated by simply putting one's money in the bank and hopping on the train with your checkbook. In this way, the advent of checks enabled much of the westward expansion that occurred in the United States throughout the 19th and early 20th centuries.

More recently, check processing was made more efficient with the passage of the Check Clearing for the 21st Century Act (Check 21), which took effect in 2004. Check 21 enabled financial institutions to accept electronic image presentment of checks in lieu of the actual physical check being presented. It further required financial institutions to accept printed reproductions of the check in lieu of the original check. The result has been that nearly all check presentment in the United States now occurs electronically.[19]

Just as we did with coins and currency, let's take a look at the check through our media attributes, as established by Poe.

Checks

Attribute	Definition	Assessment
Accessibility	the availability of the medium itself	Low, increasing to Medium: When checks emerged, the ability to read and write were very limited. Over time, increasing percentages of people gained this ability, and it would be a stretch to argue that this is a significant hindrance for many Americans at this point. That said, checks require an account at a financial institution, and around 5% of U.S. households remain unbanked.[20]
Privacy	the covertness with which data can be transmitted in a medium	Medium: Unlike coins and currency, which can be exchanged in secret between two people, the use of checks requires that at least one additional entity (and typically at least two, assuming the parties have different banks) become knowledgeable as to the amount and parties involved in the transaction.
Fidelity	the faithfulness with which data can be transmitted in a medium	Low: While checks are very limited in this regard, they do result in a record of the parties, amount, and potentially a memo line entry that is part of the payment device, leaving some amount of information related to the nature of the transaction.

Volume	the quantity in which data can be transmitted in a medium	High: With the advent of Check 21 and the ubiquitous availability of electronic presentment, checks can be processed at astonishing volume today.[21]
Velocity	the speed with which data can be transmitted in a medium	Low: Bear in mind that this is in today's context. While checks might be faster than cash in some ways, the fact is that the fastest true check transactions (meaning those that are not converted to Automated Clearing House (ACH) are agonizingly slow by today's standards.
Persistence	the duration over which data can be preserved in a medium	High: Check images can be linked to a user's account, and preserved in that way perpetually.
Searchability	the efficiency with which data can be found in a medium	Low: because checks all result in a transaction in the user's checking account (always on the payer side, and often on the payee side), that creates a record that can be searched. The nature of that search, which might be a manual review of images of all checks drawn on an account, leaves much to be desired, though.

The next major innovation in bank account based payments (and please note that now we are going to narrow the aperture to focus on the United States specifically) was the advent of the ACH, which began operating in the 1970s.[22] The ACH network was created when bankers realized they didn't necessarily need the entire check to post transactions. What they needed was the MICR (magnetic ink character recognition) information and the amount of the check.[23]

It's difficult to put into words how transformative the ACH network has been for the operation of the financial system in the United States. This innovation opened up the capability to post payroll to accounts within a day or two, compared to printing a check, getting that check in the hands of the employee (and hoping the employee won't misplace it or otherwise render it unusable), who then had to deposit it with their financial institution, which in turn needed to present the check to the employer's financial institution, which would result in settling the transaction and the funds arriving in the employee's account.

ACH has, of course, also been the underlying payment technology that underpins so many of the fintech wallets that have come to permeate our lives. These networks leverage the ubiquitous, reliable, and low-cost nature of the ACH network to facilitate 21st century experiences built on 1970s technology.

ACH

Attribute	Definition	Assessment
Accessibility	the availability of the medium itself	Medium: while essentially all accounts in the United States are connected to the ACH network, only financial institutions can input transactions into the network, meaning others who want to use the network must do so through a financial institution.

Privacy	the covertness with which data can be transmitted in a medium	Medium: Similar to checks, transactions conducted through the ACH necessarily involve at least a handful of intermediaries, all of whom become privy to some aspects of the value exchange.
Fidelity	the faithfulness with which data can be transmitted in a medium	Medium: While the ACH messaging certainly enables the exchange of far more detailed transaction information than is possible using a check or cash, there are also limitations to the amount and type of information that can be associated with an ACH payment.
Volume	the quantity in which data can be transmitted in a medium	High: The ACH network moved 30 billion transactions, accounting for the exchange of more than $72 trillion in 2022.[24] Those are big numbers.
Velocity	the speed with which data can be transmitted in a medium	Medium: Now we are getting somewhere. Classic ACH certainly leaves much to be desired in terms of speed, but in the last decade we've seen the advent of same day ACH. Moving to same day clearing and settlement certainly makes a big difference for a great many use cases.
Persistence	the duration over which data can be preserved in a medium	High: ACH transactions are posted to the user's account, and the record of those transactions can be preserved in that manner perpetually.

Searchability	the efficiency with which data can be found in a medium	Low: Again, because all ACH transactions result in a posting to the associated DDAs, they are theoretically searchable. That said, much of the addenda information (which are additionally records that can be recorded in the ACH system) isn't easily accessible by the account holder, meaning much of the original transaction is functionally lost if it is not stored in some other medium, where it is not as easily linked back to the transaction.

Before we wrap up this section, I wanted to revisit our timeline, as promised:

- First 175,000 Years: Speech only

- 5,000 Years Ago: Written Word Emerges

- 2,600 Years Ago: Lydians introduce the first coins

- 1,000 Years Ago: Earliest forms of checks, known as "saaks," emerged

- 500 Years Ago: Printed Word Emerges

- 50-60 Years Ago: ACH begins operating

Section 3 – The Vast Potential of Faster Payments

"Faster payments is the truth."[25]

- Wole Coaxum – CEO, MoCaFi

In section 2 we looked at how various advancements in payments technology have enabled massive new flows of commerce. This point does not undermine the idea that these technologies were pulled into existence and usage by demand. To the contrary, what we saw in Section 2 of this chapter was that existing needs pulled these new payments media into prevalence and then once they were out in the marketplace, innovators found new, exciting ways to combine the payments media with various other innovations to create wholly new lines of business.

And it's that phenomenon that I am hoping to scratch the surface of here briefly before getting into a massive opportunity – to better serve vulnerable populations who today are disproportionately likely to be unbanked or underbanked. Before we get into that essential line of thought, though, I want to zoom out and think about how new technologies and this new payments media can work in tandem to be true game changers.

As I discussed above in Section 1 of this chapter, we have a blueprint of sorts for this. The most recent example, which is discussed above, was the combination of payment card technology with the internet to enable e-commerce and eventually mobile commerce. It's now been years since the percentage of card transactions conducted via card-not-present flows surpassed the card-present (in-person) flows. That would have been unimaginable 30 years ago. Now it's hardly even surprising. That's how powerful the combination of modern payments media and new, seemingly unrelated, technological advances can be.

This isn't a book about Artificial Intelligence (AI), or Application Programming Interfaces (APIs), or chatbots, or QR codes, or whatever other new technology might be out there. It's a book about payments, and hopefully by now you agree that means it is also a book about communication. So what happens when you combine instantaneous media, the messages associated with which are inherently valuable, with powerful emerging technologies such as artificial intelligence?

I can tell you that I don't know the answer to that question. If I did, I'd probably be rolling in VC money. Consider someone attempting to answer that question about the convergence of card payments and the internet in the early 1990s. Few people could see the truly transformational possibilities presented by those enabling technologies, and those few people started companies like Amazon, Netflix, and eBay.

The fact is, there is tremendous untapped potential here, and while existing use cases have pulled instant payments out of the imagination and into the real world, those aren't the use cases that will make us look back at this in thirty years and wonder at how far we've come. But before we get to all of that, there are very real, highly impactful implications of this new technology. That's where I'm going to leave you, with a huge societal imperative underpinning all of this.

Serving the Underserved

"The law, in its majestic equality, forbids rich and the poor alike to sleep under bridges." Anatole France

You've probably seen that quote from Anatole France before, and hopefully I don't need to explain what's meant there. What's that got to do with payments, though?

Just as the law forbids rich and poor to sleep under bridges, the banking system in the United States welcomes anyone who wants a bank account. The primary difference between the two? Generally speaking, more wealthy households are the ones who want bank accounts, at least historically.

Just to level set on that point, which no reader should take for granted, in a 2021 study the Federal Deposit Insurance Corporation (FDIC) found that 4.5% of U.S. households were unbanked, and found the unbanked rates were higher among lower income households, Black households, Hispanic households, and single-mother households. Further, the FDIC found that 14.1% of households were "underbanked" – that's nearly 19 million *households*. For comparison, the New York Metropolitan Statistical Area has about 19 million *people*.

To put these disparities into context, the FDIC also found that 81.6% of fully banked households exclusively used their bank account to pay bills or receive income while only 38.1% of underbanked did so. Additionally, the use of mobile banking is *higher* among underbanked households (48.8%) than among fully banked households (42.5%). What's going on here?

Well, there are probably lots of things going on, but one thing seems pretty clear. There are plenty of people who have *access* (in the broad sense) to bank accounts and who are choosing not to have them, or not to use them. Why would that be? Well, just because individuals can access a service does not mean that service is *inclusive* of them in the sense that it works how they want it to work and does the things they need it to do. Living on the financial edge – being one misstep away from a potentially catastrophic cascade of events presents different needs than those experienced by higher-income households.

So what are these pain points that keep people from being banked or from using a bank account in the ways higher-income households use theirs?

The U.S. Faster Payments Council's (FPC) Financial Inclusion Work group helps us answer this question, so let's look at the pain points they've identified:

- Design: product design not targeted to the needs of the financial lives of the underserved

- Liquidity Constraints: tight budgets mean that a delay, interruption, or loss of funds can lead to a cascade of adverse financial consequences

- Cash in/Cash out: people without a bank or credit union account face costs to get cash into and out the digital economy

- Trust: Trust in financial services providers may be low. It may be undermined by the absence of strong customer service and language access (recall that Hispanic households are disproportionately likely to be unbanked. For that reason, it's worth noting that according to the U.S. Department of Health and Human Services Office of Minority Health, about 28% of Hispanic Americans do not speak English well), and by apprehension.

- Mistake prevention: limited ability of the customer to absorb the loss of funds due to a mistake.

- Fraud prevention and remedy: limited ability of the customer to absorb the loss of funds from fraud.

- Security: Concerns about the security of funds and the impact of security procedures on inclusion.

- Interoperability for ease of use: Concerns about how to use faster payments efficiently when there are many systems and payees and payors may use different systems that do not connect.[26]

It isn't about simply being invited to participate. It's about being offered a set of products and services that meet your unique needs. People who are living paycheck to paycheck have a different set of needs than people who have a bit more cushion.

One statistic we've all likely heard in one form or another is that nearly six in 10 Americans don't have enough savings to cover an unexpected $500 expense.[27] But what if the payments system works in such a way as to impose *effective* temporary costs on its users in the form of funds being trapped in transit? While this is a temporary inconvenience for many of us, for those who live close to the financial edge, the inability to access all of their funds that are in transit can lead to a cascade of late payments (resulting in late fees, re-connection fees, overdraft fees, and so forth).

This is what's meant by "Liquidity Constraints" above, and it's one of the principal ways that faster payments can make being banked (or using a bank account) more attractive. This is because instant payments can remove the variables presented by the use of various payments systems, which settle and clear transactions at differing and (at least to consumers) unpredictable rates. Instant payments offer an experience that is intuitive and predictable. And, when you consider that cashing a check in the United States costs, on average, 2.34% of the face value of the check ($12 for a $500 check), providing an electronic option that is simple, quick, and less expensive should be awfully attractive to a large portion of today's unbanked population.

The trouble with cash, though, if we look at the various attributes of communication media, is that its velocity is only real-time for face-to-face interactions which means the range of cash as a real-time payment method is literally as long as your arm. Sure, cash can be sent over longer distances, say by sending it in the mail (which your parents told you never to do, but now what shows up in envelopes addressed to your kids from those same parents?).

Beyond liquidity management, though, is another pain point that keeps people from deciding to be banked. And that is trust. There are clearly a number of dimensions to trust, and I note that faster payments will not resolve all of the issues that might result in a gap between would-be customers and financial institutions. But when it comes to the questions about incurring unexpected overdraft fees or having the liquidity you need available when you need it, transacting in real time can make a real difference in building trust.

In short, the advent and provision of instant payments systems can dramatically increase the attractiveness of developing and using banking relationships. This is often the first step to financial gains in terms of savings, building credit, and other milestones to which we all strive. Instant payments provide the promise that these things can be in reach for more and more Americans. Unlike so many other technologies, which have had a tendency to leave certain vulnerable populations behind, instant payments can truly be the tide that lifts all boats.

Ultimately, though, the ability to bring more people into the fold, and ultimately to serve those people with financial services that are relevant to their needs, is only scratching the surface of how transformative instant payments can be.

Conclusion

Wrapping Up

As we bring this chapter to a close, let's call back to those tables I used as we flipped through the history of payments as a communication methodology and take a look at how instant payments systems stack up.

Attribute	Definition	Assessment
Accessibility	the availability of the medium itself	Currently Low, but increasing quickly: At the time of writing, about 65% of DDAs in the United States are able to receive an instant payment. Far fewer are capable of sending. Also, as is the case with check and ACH, instant payments require an account at a financial institution, and around 5% of U.S. households remain unbanked.[28]
Privacy	the covertness with which data can be transmitted in a medium	Medium: Similar to checks and ACH, transactions conducted through instant payments networks necessarily involve at least a handful of intermediaries, all of whom become privy to some aspects of the value exchange.

Fidelity	the faithfulness with which data can be transmitted in a medium	High: The use of ISO 20022 messaging capabilities enables considerable transfer of value and associated data. Additionally, the availability of the request for payment functionality creates opportunities for in-context communications related to payments. Finally, the ability to attach documents, such as PDFs of invoices, to the payment message, makes the communication aspect of instant payments extremely robust.
Volume	the quantity in which data can be transmitted in a medium	Low, but growing quickly. At the time of writing, The Clearing House had recently announced its first-ever 1 million transaction month for the RTP network. Many forecasters are estimating that the RTP network will process more than 500 million transactions in 2024.
Velocity	the speed with which data can be transmitted in a medium	High: The name says it! Final clearance and settlement within seconds.
Persistence	the duration over which data can be preserved in a medium	High: Instant Payment transactions are posted to the user's account, and the record of those transactions can be preserved in that manner perpetually.
Searchability	the efficiency with which data can be found in a medium	High: Because these networks are established with communication in mind, and because they are intended to serve in that regard, I expect that the user interfaces will enable a high degree of searchability related to the messages and documents that are exchanged in conjunction with payments.

If nothing else, this should demonstrate the tremendous potential this new payments technology holds. We've established that payments are another medium for communication, an assertion previously made by Swartz in her book, to be sure. We've also discussed at some length the fact that new media are not brought into being by the availability of new technology.

People did not begin writing because they suddenly had the ability to do so. The printing press wasn't invented by Gutenberg because it suddenly became possible (it wasn't invented by Gutenberg at all, but that's a story for a different book, the Poe book I cited repeatedly throughout this chapter). The same goes for broadcast media, and for the internet. They didn't come into being or achieve scale because they suddenly became possible. Instead, they were "pulled" into existence and widespread use by demand.

The same is the case for instant payments. So where are these sources of demand? Much ink has been spilled, and many pundits have made their names discussing instant payments use cases, and I absolutely love talking about the use cases for instant payments (though at the end of the day, in my view, instant payments use cases are any legal reason why one party wants to send money to another party). But in some ways, zooming in on specific use cases misses the bigger picture – this is a foundational technology that has the potential to enhance a myriad of use cases and enable the creation of new, innovative business models that weren't previously even possible.

From the potential for micro-payments to revolutionize how we consume content (and potentially how we are paid!), to innovative methods of lending, to untold combinations of new technologies like AI with payments. We are on the cusp of something that will transform the way we live our lives. Let's see what happens.

Chapter 2 - History of Instant Payments

By Peter Davey & Connie Theien

A Brief History of Payments

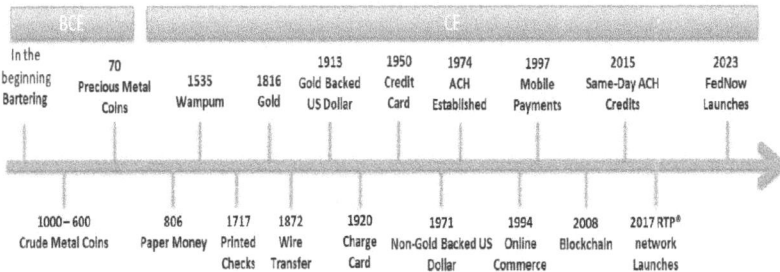

BCE				CE					
In the beginning Bartering	70 Precious Metal Coins	1535 Wampum	1816 Gold	1913 Gold Backed US Dollar	1950 Credit Card	1974 ACH Established	1997 Mobile Payments	2015 Same-Day ACH Credits	2023 FedNow Launches
1000–600 Crude Metal Coins	806 Paper Money	1717 Printed Checks	1872 Wire Transfer	1920 Charge Card	1971 Non-Gold Backed US Dollar	1994 Online Commerce	2008 Blockchain	2017 RTP® network Launches	

The concept of payments (the transfer of money, goods, or services in exchange for goods and services[1]) has been around for a long time dating back to the early days of the bartering system in Mesopotamia around 6000 BC. It wasn't until 1872 when Western Union launched the Electronic Fund Transfer (EFT) system on top of telegraph technology that we were able to start to envision how we would move money electronically.

Even with the advent of the first EFT system, it wasn't until the 1950's that financial institutions started to use computer systems to automate back-office tasks. Bank of America was credited as one of the first to leverage computer systems in the 1950s to process check transactions using magnetic ink character recognition (MICR). [2]

The true concept of electronic retail payments didn't materialize until the late 1960s. It was through the work of the Special Committee on Paperless Entries (SCOPE) and The American Bankers Association who both were looking at ways of improving payment systems in the United States. The work by these groups led to the formation of the first Automated Clearing House (ACH) association in California in 1972 and the formation in 1974 of the Electronic Payments Association, which became the National Automated Clearinghouse Association (Nacha) in 1985.[3]

While the advent of ACH in the 1970s was a step that enabled the industry to take previously paper-based processes and start to electrify them, the larger more long-term results of this movement paved the way for broader adoption of electronic payment processing. This resulted in advancements and adoption of new technologies in banking combined with the growth of connected systems through the adoption of the internet by both financial institutions and their customers.

The Anatomy of a Payment

Before we get to faster and instant payments we need to make sure we are all speaking the same language. The following anatomy of a payment should be helpful.

INITIATION	The initiation of a payment begins when either the payer or payee in a payment transaction, or a third party, sends an instruction to another entity that triggers a process ultimately leading to a payment.
AUTHENTICATION	The process that verifies the identity or veracity of a participant, device, payment or message connected to a payment system. Authentication may happen at multiple points in the payment process. For example: •End-user identity may be verified when the end user enrolls with a provider. •During the payment process, additional checks may be built in to verify the identity of the payer, account, or account provider (e.g., entering a password).
AUTHORIZATION	The explicit instructions, including timing, amount, payee, source of funds and other conditions given by the payer to their account provider or to the payee to transfer funds on a one-time or recurring basis.
APPROVAL BY THE PAYER'S PROVIDER	The point following the initiation of a payment when the payer's account provider verifies that the payer's account has good funds[3] or credit necessary to complete the transaction.
CLEARING	The process by which the payer's and payee's account providers exchange payment information to confirm a transaction prior to settlement.
RECEIPT	The point when funds are received by the payee, such that the funds can be withdrawn or transferred.[4]
SETTLEMENT	An act that discharges obligations in respect of funds between two or more entities.[5]
RECONCILIATION	A procedure to verify that the records issued by entities involved in a transaction match. The reconciliation process may include appropriate reversals and post-transaction analysis.

19

Industry Faster Payment Initiatives

On the heels of the Check 21 Act implementation in 2004, which transformed and accelerated check clearing through exchange of images, the industry turned its attention to accelerating clearing and settlement of retail payments more broadly. Initial focus was on developing the option for same-day settlement of ACH transactions, followed by industry efforts to develop solutions including Push-to-Card and Instant payments that would provide immediate clearing and availability of funds.. We will proceed to take a look at each of these, with the deepest dive into those related to instant payments.

» Same-Day ACH

While ACH continued to grow in adoption and new features, there hadn't been significant improvements to the speed in which money settled since the launch of the ACH Network. All of that changed in 2010, when the Federal Reserve launched the FedACH SameDay Service that allowed for same-day settlement of ACH batch files. This was limited to transactions arising from consumer checks converted to ACH and consumer debit transfers initiated over the Internet and phone.[6]

Initially an opt-in service, same-day ACH didn't see broad adoption until the industry rallied around a set of rules changes for the ACH network that mandated receipt of same-day transactions and culminated in the launch of same-day ACH credits in September of 2016. The new rules required receiving depository financial institutions (RDFIs) to participate in the service and originating depository financial institutions (ODFIs) to pay a fee to RDFIs for each same-day ACH forward transaction.[7]

Same-day ACH, combined with later processing deadlines and other enhancements, has spurred continued volume growth in the ACH network. That said, it hasn't addressed industry and end-user demand for instant payment solutions.

» Push-to-card - 2011-2015 Visa Direct and Mastercard Send

Visa and Mastercard both began development of capabilities that would enable merchants to leverage the debit card rails to "push" a payment to a customer with a 16-digit Primary Account Number (PAN). This work started with Visa introducing the Original Credit Transaction (OCT) around 2011 but push-to-card really didn't start gaining adoption until around 2015 with the launch of Visa Direct and Mastercard Send.

These advancements on the card networks sped up delivery of transaction clearing and funds availability for end customers. The settlement between financial institutions remained the same as for debit card transactions, occurring 1-3 days after the transaction was debited from the payor's account and credited to the receiver's account.

In terms of perceived instant payment capabilities for consumers, push to card has gained traction as an attractive option for commercial entities and fintechs wanting to create an "instant payment" experience for customers. Although the lack of immediate and final settlement creates credit risk, the option benefits from established, ubiquitous and accessible infrastructure.

» Federal Reserve Payment System Improvement Efforts - 2012 Proposed Vision

In October 2012, the Fed shared their vision for potential payment system improvement in a speech given by Sandra Pianalto, President of the Federal Reserve Bank of Cleveland and Chair of the Federal Reserve's Financial Services Policy Committee. The heart of the vision was to improve the speed and efficiency of the U.S. payment system from end-to-end over the next decade while maintaining a high level of safety and accessibility. End-to-end meant addressing the needs of and engaging end users and other stakeholders in the payment system versus previous efforts focused on primarily on interbank improvements. The Fed committed to the continual improvement of the U.S. payments system and to working collaboratively with the payments industry to implement innovations that meet evolving payment needs.[8]

» Federal Reserve Payment System Improvement Efforts – 2013 Public Consultation Paper

In September of 2013, the Federal Reserve Banks published the "Payment System Improvement" public consultation paper [4] that sought to engage industry stakeholders in discussion about how the U.S. payments system should evolve to meet the needs of the future. This paper acknowledged that other countries were adopting new ways of doing payments and that there were some new payment services and technologies emerging in the US as well.

New electronic networks were proliferating, including networks for person-to-person transfers, online merchants, business trade payments, and others. However, many of these networks were "closed loop" systems making it inconvenient or impossible for in network end users to make or receive payments to or from out of network end users. By contrast, "open loop" legacy payment systems were nearly ubiquitous and allowed end users to send payments to almost any receiver, without requiring the receiver to enroll in the system to retrieve the payment.[4]

The paper noted deficiencies or gaps in the payment system that potentially needed to be addressed, including:[4]

- Slowness in availability of data and funds to end users

- Continued reliance on checks

- Exposure of sensitive account information required to initiate payments

- Lack of transparency and timely notifications

- ◉ Lack of ubiquity in emerging innovations

- ◉ Increasing security and fraud threats

- ◉ Risks and lack of certainty inherent in deferred settlement solutions

- ◉ End user reluctance to adopt electronic payments

- ◉ Lack of convenient, cost-effective, and timely cross-border payment solutions

The paper summarized the gaps and opportunities in the following statement.

"End users of payment services are increasingly demanding real-time transactional and informational features with global commerce capabilities. Legacy payment systems provide a solid foundation for payment services; however, some of these systems (e.g., check and ACH) rely on paper based and/or batch processes, which are not universally fast or efficient from an end user perspective by today's standards. The challenge for the industry is to provide a payment system for the future that combines the valued attributes of legacy payment methods – convenience, safety, and universal reach at low cost to the end user – with new technology that enables faster processing, enhanced convenience, and the extraction and use of valuable information that accompanies payments".

There was overwhelming feedback to the consultation paper from stakeholders across industry including over 200 responses from non-bank entities. Responses were submitted by individual financial institutions, businesses, payment networks and processors, software vendors, payment innovators, consultants, and consumers, as well as from trade groups representing financial institutions and business members. [9]

» Federal Reserve Payment System Improvement – 2015 Strategies for Improving the U.S. Payment System [10]

Incorporating input and insights from the public consultation paper, multiple research efforts and an extensive stakeholder engagement program, the Federal Reserve released in January 2015 a comprehensive report titled "Strategies for Improving the U.S. Payment System." [10] This report became the basis for improvements in the U.S. payment system, as it laid out a strategic roadmap for enhancing the efficiency, security, and speed of payments in the United States.

The objective of this report was to address the shortcomings of the existing payment system and to provide a framework for modernization with the following industry desired outcomes in mind.

- **Speed:** A ubiquitous, safe, faster electronic solution(s) for making a broad variety of business and personal payments, supported by a flexible and cost-effective means for payment clearing and settlement groups to settle their positions rapidly and with finality.[10]

- **Security:** U.S. payment system security that remains very strong, with public confidence that remains high, and protections and incident response that keeps pace with the rapidly evolving and expanding threat environment.[10]

- **Efficiency:** Greater proportion of payments originated and received electronically to reduce the average end-to-end (societal) costs of payment transactions and enable innovative payment services that deliver improved value to consumers and businesses.[10]

- **International:** Better choices for U.S. consumers and businesses to send and receive convenient, cost effective and timely cross-border payments.[10]

- **Collaboration:** Needed payment system improvements are collectively identified and embraced by a broad array of payment participants, with material progress in implementing them.[10]

The paper identified research, policy development and industry collaboration efforts to be facilitated by the Federal Reserve in pursuit of the identified desired outcomes. Specific to the desired outcome for payment system speed, the Federal Reserve conveyed its intent to establish and lead an industry Faster Payments Task Force charged with assessing options for achieving the desired capabilities for the United States.

» Federal Reserve Payment System Improvement - 2015-2017 Faster Payments Task Force

The Federal Reserve convened the Faster Payments Task Force in May 2015 with membership of 320 industry stakeholders representing financial institutions, technology providers, merchants, consumer groups, trade groups and regulators.[13]

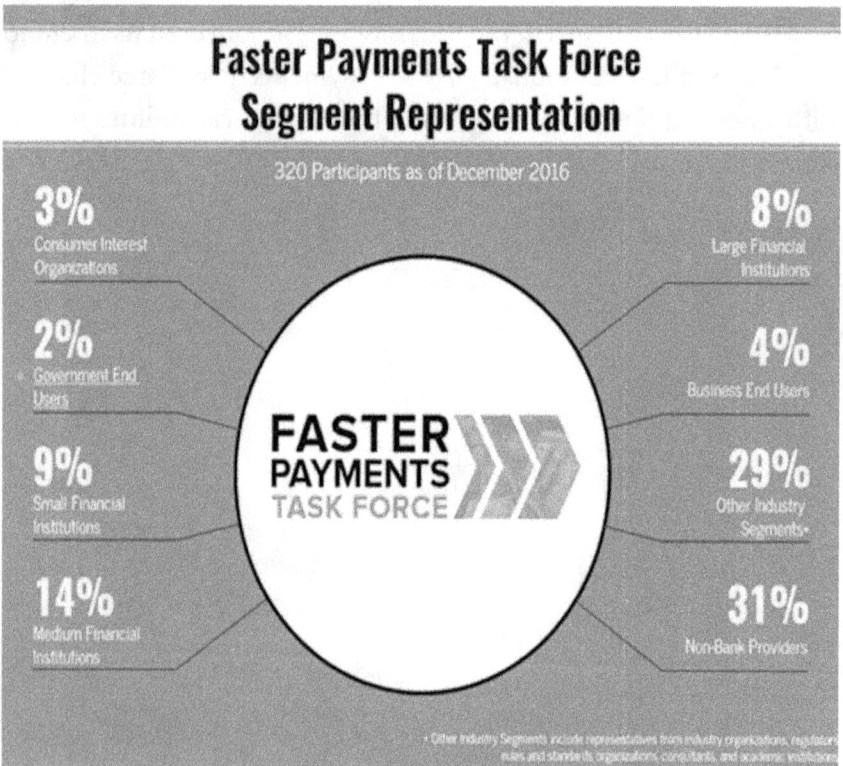

Faster Payments Task Force Segment Representation

320 Participants as of December 2016

3% Consumer Interest Organizations

2% Government End Users

9% Small Financial Institutions

14% Medium Financial Institutions

8% Large Financial Institutions

4% Business End Users

29% Other Industry Segments•

31% Non-Bank Providers

• Other Industry Segments include representatives from industry organizations, regulatory rules and standards organizations, consultants, and academic institutions

12

Steering committee members were elected by the membership from each stakeholder category within the task force. Three appointments were made by the Federal Reserve task force chair to ensure that stakeholder perspectives were sufficiently represented.[13]

The work of the Faster Payment Task Force comprised five phases designed to align the industry around desired attributes for future systems and explore possible solutions, challenges and opportunities to support progress.

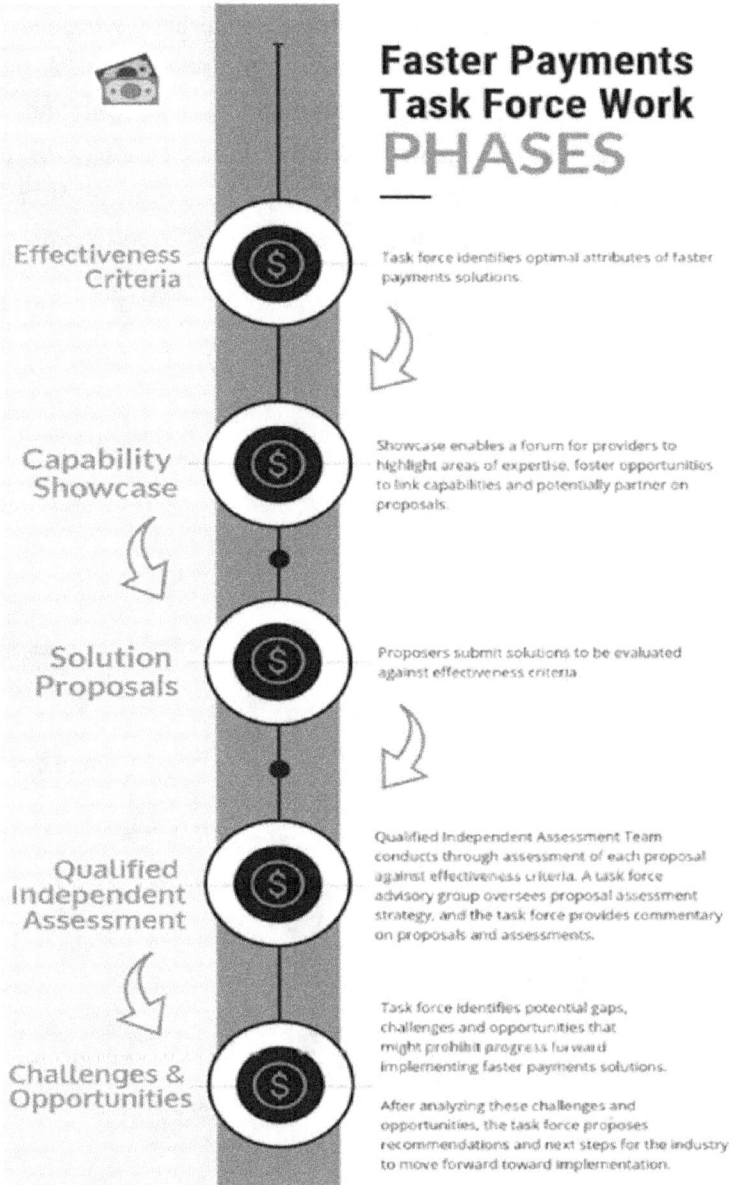

Faster Payments Task Force Work
PHASES

Effectiveness Criteria — Task force identifies optimal attributes of faster payments solutions.

Capability Showcase — Showcase enables a forum for providers to highlight areas of expertise, foster opportunities to link capabilities and potentially partner on proposals.

Solution Proposals — Proposers submit solutions to be evaluated against effectiveness criteria.

Qualified Independent Assessment — Qualified Independent Assessment Team conducts through assessment of each proposal against effectiveness criteria. A task force advisory group oversees proposal assessment strategy, and the task force provides commentary on proposals and assessments.

Challenges & Opportunities — Task force identifies potential gaps, challenges and opportunities that might prohibit progress forward implementing faster payments solutions.

After analyzing these challenges and opportunities, the task force proposes recommendations and next steps for the industry to move forward toward implementation.

14

» Federal Reserve Payment System Improvement - 2015-2016 Instant Payments Effectiveness Criteria

The first, and one of the most impactful, accomplishments of the Faster Payments Task Force was development and publication of the "Faster Payments Effectiveness Criteria " which outlined 36 attributes of an effective faster payment system along with descriptions and a rating scale for each attribute. The purpose was to guide development of industry solutions in alignment with defined stakeholder needs and to create criteria by which solution alternatives could be evaluated. The effectiveness criteria were organized into the following 6 categories: [16]

UBIQUITY	U.1	Accessibility
	U.2	Usability
	U.3	Predictability
	U.4	Contextual Data Capability
	U.5	Cross-Border Functionality
	U.6	Applicability to Multiple Use Cases
EFFICIENCY	E.1	Enables Competition
	E.2	Capability to Enable Value-Added Services
	E.3	Implementation Timeline
	E.4	Payment Format Standards
	E.5	Comprehensiveness
	E.6	Scalability and Adaptability
	E.7	Exceptions and Investigations Process
SAFETY AND SECURITY	S.1	Risk Management
	S.2	Payer Authorization
	S.3	Payment Finality
	S.4	Settlement Approach
	S.5	Handling Disputed Payments
	S.6	Fraud Information Sharing
	S.7	Security Controls
	S.8	Resiliency
	S.9	End-User Data Protection
	S.10	End-User/Provider Authentication
	S.11	Participation Requirements
SPEED (FAST)	F.1	Fast Approval
	F.2	Fast Clearing
	F.3	Fast Availability of Good Funds to Payee
	F.4	Fast Settlement Among Depository Institutions and Regulated Non-Bank Account Providers
	F.5	Prompt Visibility of Payment Status
LEGAL	L.1	Legal Framework
	L.2	Payment System Rules
	L.3	Consumer Protections
	L.4	Data Privacy
	L.5	Intellectual Property
GOVERNANCE	G.1	Effective Governance
	G.2	Inclusive Governance

19

The phases of task force work solicited task force members to showcase relevant marketplace capabilities that could be part of an effective solution, and subsequently, to bring forward comprehensive faster payment solution proposals to be evaluated and published to the industry, with the objective of spurring innovation and catalyzing awareness and adoption of private sector offerings. An independent consultant was retained to evaluate the 24 submitted solution proposals against the effectiveness criteria. Following is a listing of the final 16 proposals that ended up being documented in the final task force report.

FASTER PAYMENTS SOLUTION PROPOSALS

Proposers	Solution Summary
DWOLLA, INC.	**Description:** Real-time transfer solution features 24x7x365 credit-push operations, defines responsibilities by participant, is interoperable with other solutions. **Scope:** Comprehensive end-to-end faster payment solution that provides models for directory, application program interface (API) accessibility, and variable credit push requests. **Approach:** Facilitates competition through modern means of access (i.e., APIs) and end-user ability to choose among providers, use multiple accounts, and specify payment preferences.
HUB CULTURE, ECCHO, XALGORITHMS	**Description:** Distributed Ledger Technology like solution with a Real-time Asset Interchange Network (RAIN) and Real-time Asset Interchange Ledger (RAIL) to enable end-to-end global payment transactions. **Scope:** Comprehensive end-to-end faster payment solution. **Approach:** Includes a Unique Synchronized Identity (USI) to link existing accounts with the potential to lower transaction costs and platform outlays and to reduce time to deploy across the network supporting push and pull payments.

Proposers	Solution Summary
INTERCOMPUTER CORPORATION	**Description:** Real-time payments system with multi-layer security features messaged on private internet channel without using Web protocols. Utilizes expedient 3-factor authentication. Payments fully insured by underwriting. **Scope:** Comprehensive, end-to-end, hosted/ distributed payment solution for all computing devices eliminating web cybercrime. **Approach:** Allows for users to set up a hierarchy of accounts and uses APIs to interface with other customer financial systems (e.g., treasury management).
KALYPTON GROUP LIMITED AND ECCHO	**Description:** Full transaction processing engine, not just payments, delivering blockchain-like functionality without its limitations, through distributed authentication of private ledgers for real-time account-to-account transactions. **Scope:** Comprehensive end-to-end faster payments solution that supports both bank and nonbank providers. **Approach:** Incorporates a configurable service layer that implements multiple use cases and supports rapid, ongoing innovation by service providers.

Proposers	Solution Summary
MOBILE MONEY CORP.	**Description:** Cash-based, closed-loop solution using MoMo accounts to process banking transactions including faster payments. **Scope:** Targeted to serve the unbanked, underbanked and some banked. **Approach:** Relies on network of bank and nonbank agents that register end users and provide cash-in and cash-out capabilities.
NANOPAY CORPORATION	**Description:** Direct, collateralized bearer asset transfer system that enables immediate settlement when sending money and other assets between users. **Scope:** Closed-loop faster payment system using digital currency with easy integration via ISO 20022 and open APIs. **Approach:** Enables end users to switch among providers enabling competition, has redundancy in operating design.

Proposers	Solution Summary
NORTH AMERICAN BANKING COMPANY AND INDEPENDENT COMMUNITY BANKERS OF AMERICA	**Description:** Provides front-end to the ACH leveraging same-day credit push for payment transactions through their All Payments App and a Directory for mapping to individual bank accounts.
	Scope: Solution facilitates same-day transactions through the ACH network.
	Approach: Immediately achieves ubiquity (among banks) and is cost effective through use of existing ACH platforms, standards, rules and record formats.
RIPPLE	**Description:** Leverages distributed financial technology to enable real-time cross-border payments for financial institutions that join their network.
	Scope: Uses Interledger to enable real-time cross-border payments and interoperability between systems.
	Approach: Allows for originators to see total cost of payment and real-time tracking for low- and high-value payments.

Proposers	Solution Summary
SHAZAM, INC.	**Description:** Develops case for universal sharing of information and equitable participation among debit networks and creates a governance model to process faster payments. **Scope:** Enables movement of debit or credit message transactions between bank deposit accounts in real time. **Approach:** Achieves ubiquity by leveraging the debit card networks and eliminates need for a directory by routing transactions through account token prefixes.
SWAPSTECH	**Description:** Enables individuals, businesses and merchants to use ZillPay, a domestic and global payment network, to make secure, fast and convenient payments, using alias directory. **Scope:** Addresses practice of current payment networks and financial systems allowing account numbers to be shared with third parties, leaving the accounts open to fraudsters. **Approach:** Uses universally accepted payment address, UPA, to ensure the account remains private, and there is no sharing of account numbers for any payment.

Proposers	Solution Summary
THE CLEARING HOUSE AND FIS	**Description:** A real-time payment solution for all financial institutions based on internationally tested solutions and established structures to enable account-to-account real-time credit push payments at all U.S. depository financial institutions. **Scope:** Comprehensive end-to-end faster payment solution. **Approach:** Operates 24x7x365 and leverages the safety and security features of existing bank channels and risk management controls.
THOUGHT MATRIX CONSULTING, LLC	**Description:** A central bank issued electronic currency, Money Modules (MM), based on blockchain technology that features standardized software, data modules and electronic wallets. **Scope:** Closed-loop faster payment system using digital currency. **Approach:** Addresses all use cases including the unbanked and offers customizable and configurable features for providers and end users.

Proposers	Solution Summary
TOKEN, INC.	**Description:** Provides a set of software and service components that enables payers to make direct push and pre-authorized pull payments using digital tokens for depository institutions and nonbank financial service providers. **Scope:** Comprehensive end-to-end faster payment solution. **Approach:** Uses security that protects end-user data and reduces payment risk with digital signatures and certificates to authenticate and authorize each transaction.
UNIVERSITY BANK	**Description:** PayThat is a closed loop real-time payment system providing superior privacy and security while significantly lowering transaction costs. **Scope:** Comprehensive "real-time" highly secure payment system that interoperates with existing bank systems. **Approach:** Leverages email and texting to send and receive good-funds transactions and describes four business models of detailed value added services to speed adoption and enhance the Faster Payments business case.

Proposers	Solution Summary
WINGCASH	**Description:** A software platform that would be owned and operated by the Federal Reserve and the Governing Organization. The Federal Reserve would issue digital currency (digital Fed notes) and is tied to the Internet domain (Fednotes.com). **Scope:** Faster payment system using Federal Reserve-issued digital currency. **Approach:** Allows businesses to issue and define the characteristics (i.e., name, redemption) of digital brand cash.
WORLD CURRENCY USA, INC.	**Description:** Allows financial institutions, using FAST solution, to connect through a web-based front-end platform and to clear/ settle payments through FAST clearing accounts. Separate platform allows users to initiate and view their transactions in real time. **Scope:** Comprehensive end-to-end faster payment solution. **Approach:** The solution proposes to consolidate domestic ACH and wire payments into a single payments solution for domestic and cross-border payments.

18

» Federal Reserve Payment System Improvement - 2017 The U.S. Path to Faster Payments

The Faster Payments Task Force concluded its work by issuing its final report, "The U.S. Path to Faster Payments" in two parts; Part One: The Faster Payments Task Force Approach in January 2017 and Part Two: A Call To Action in July 2017. Part One summarized the task force background, objectives and approach and provided foundational industry analysis that defined the problems to be addressed. Part Two summarized the go forward vision, challenges, a path forward in the form of 10 recommendations, and an industry call to action.

These reports marked a significant milestone in the efforts to modernize the U.S. payment system, offering a clear roadmap for the transformation of payment infrastructure in the country.

The recommendations and key takeaways from the paper highlighted:

- The desire for ongoing stakeholder collaboration in the form of an industry governance entity and engagement on market practices and regulation;

- Infrastructure needs, including a 24x7 Federal Reserve settlement infrastructure, industry directory interoperability, and network ubiquity, potentially advanced through Fed participation as an operator;

- The importance of end user education and a strong focus on security and fraud prevention to fuel successful adoption; and

- The need to build an adaptable ecosystem for the long term, considering future cross-border capabilities and the potential of emerging technologies.

The Faster Payments Task Force envisioned a system where solutions both compete and interoperate to provide payment services that are faster, ubiquitous, broadly inclusive, safe, highly secure and efficient. Its final report included a call to action for all payments stakeholders to: [17]

- Embrace and promote the vision and the Effectiveness Criteria;

- Actively participate in ongoing industry dialogue;

- Contribute to work group efforts and deliverables; and

- Take steps to ready their organizations for faster payments.[17]

A few months after the Faster Payments Task Force issued its final report, which called on the Federal Reserve to support several industry efforts and take action as an operator, the Fed published "Federal Reserve Next Steps in the Payments Improvement Journey" which laid out its payments improvement next steps including plans to address the recommendations from the task force report. The Fed announced its commitment to:

- Facilitate continued industry collaboration to develop a faster payments ecosystem by sponsoring three new task force-proposed work groups to design a governance framework, explore directory interoperability and assess the need for rules, standards or regulatory changes.

- Pursue Federal Reserve settlement services that address the future needs of a ubiquitous instant retail payments environment. [11]

- Explore and assess the need, if any, for Federal Reserve engagement as a service provider, beyond providing settlement services, in the faster payments ecosystem to support industry achievement of the desired outcome. [11]

- Consider other enhancements to its existing services and continue to monitor, study and solicit input from stakeholders to understand the implications of new payment technologies and models, including distributed ledger technologies and digital currencies, that can facilitate a safe and efficient U.S. payment system.[11]

» The Role of the US Faster Payments Council (FPC) - 2018

A key recommendation that came out of the GFFT was the formation of the US Faster Payments Council for the continued collaboration and efforts to provide education and awareness about faster payments for the ultimate objectives set out through all of the previous efforts leading up to this.

The Faster Payments Council (FPC) is an industry-led membership organization whose vision is a world-class payment system where Americans can safely and securely pay anyone, anywhere, at any time and with near-immediate funds availability. By design, the FPC encourages a diverse range of perspectives and is open to all stakeholders in the U.S. payment system. Guided by principles of fairness, inclusiveness, flexibility and transparency, the FPC uses collaborative, problem-solving approaches to resolve the issues that are inhibiting broad faster payments adoption in this country. [20]

As of 2023, The FPC has fostered collaboration through the following currently active work groups:

- Education & Awareness
- Directory Models
- QR Code Interface
- Financial Inclusion
- Secure Instant & Immediate Payment APIs
- Real-time Recurring Payments
- Fraud
- Cross-Border Payments
- Digital Assets in the Financial Industry
- Operational Considerations for Instant & Immediate Payments

The FPC has a resourceful "Knowledge Center" and website information, that includes whitepapers and other work group and committee deliverables. The Education & Awareness Work Group that help manage the Knowledge Center, also has the following additional valuable resources:

- Glossary of Terms (https://fasterpaymentscouncil.org/Glossary-of-Terms)

- Faster Payment Solution Providers (https://fasterpaymentscouncil.org/Service-ProvidersCatalogue)

- Faster Payment Service Providers (https://fasterpaymentscouncil.org/Service-ProvidersCatalogue)

- Use Case Repository (https://fasterpaymentscouncil.org/ use-cases)

- Faster Payments Professional Certification (https://fasterpaymentscouncil.org/Faster-Payments-Professional-Certificate)

You can also keep up with faster payment related news, events, podcasts, and blogs

Global Inventory of Instant Payment Platforms

While instant payments are relatively new in the U.S, the Faster Payments Task Force Report pointed out several other countries that had already implemented various Faster Payment Systems at that time in early 2017.

TABLE 3: GLOBAL FASTER PAYMENT SYSTEMS
SOURCE: FIS FLAVORS OF FAST 2016

COUNTRY	FASTER PAYMENT SYSTEM	YEAR OF IMPLEMENTATION
Japan	Zengin System	1973
Switzerland	Swiss Interbank Clearing—SIC	1987
Iceland	Greiðsluveitan⁵⁵	2000
South Korea	Interbank Home/Firm Banking Network—HOFINET	2001
Brazil	Funds Transfer System—SITRAF	2002
Mexico	Sistema de Pagos Electronicos Interbancarios—SPEI	2004
South Africa	Real-Time Clearing—RTC	2006
Chile	Transferencias en Linea—TEF	2008
United Kingdom	UK Faster Payments	2008
China	Internet Banking Payment System—IBPS	2010
India	Immediate Payment Service—IMPS	2010
Nigeria	NIBSS Instant Payments—NIP	2011
Poland	Express ELIXIR	2012
Sweden	Payments in Real-Time—BIR	2012
Turkey	Retail Payment System—RPS	2012
Sri Lanka	Lanka Pay	2013
Denmark	NETS Real-Time 24/7	2014
Singapore	Fast and Secure Transfers—FAST	2014
Bahrain	Fawri +	2015
Australia	New Payments Platform—NPP	Expected 2017

19

Since this time, Australia went live and numerous other countries have implemented instant payment platforms. As of the fall of 2023, there were at least 44 countries with instant payment platforms, including TCH's RTP released in 2017 and Fednow released in 2023.

Why does the US need instant payments?

Some may ask; Why do we need instant payments when there are other legacy and faster payment platforms that sufficiently work for their purposes? Instant payments systems do not necessarily replace other payment systems. They provide additional capabilities built to help US Financial Institutions and their customers participate and compete in a 21st century global economy.

Some of the differentiations and justifications for instant payments systems include:

- Simultaneous 24x7x365 instant and final funds settlement between end users and the financial institutions in seconds, providing greater control for users and reducing credit risk for financial institutions and intermediaries.

- Instant funds availability that cannot be reversed or otherwise pulled back through chargebacks

- Instant validation of transacting accounts and funds available to transfer

- Payer controlled and initiated credit-push transfers (i.e. no debit-pulled payments)

- Future international Interoperability with other global instant payment systems

- Transactions done through bank/credit union accounts which have federal deposit insurance coverage versus through nonbank alternatives

- Processing efficiency through elimination of the reporting and reconciliation activities associated with deferred net settlement alternatives leveraging the rich data and functional capabilities of ISO 20022

- Instant payments can address pain points of specific use cases and provide additional benefits as pointed out in other chapters in this book.

RTP

In October of 2014 after participating in the Federal Reserve's Public Consultation paper and undertaking its own assessment of the US payments landscape, The Clearing House (TCH) and its member banks announced that it would undertake a multi-year collaborative effort to build an instant payment system for all US financial institutions.

Vocalink, a Mastercard company, is the technology behind RTP. It was an "evolution" of its system developed for the UK, Singapore and Thailand. TCH signed a letter of intent with Vocalink in 2015 to help build and deliver core elements of RTP. [21]

RTP was designed with the Faster Payments Task Force's Effectiveness Criteria in mind, and was one of the solutions evaluated by the Faster Payments Task Force. After working with its owner banks and industry stakeholders, TCH launched their RTP® system in November 2017.

As of July 2023, RTP surpassed half a billion payments since inception, recognizing a daily volume over 1-million payments in September 2023. The increasing volume on the RTP network was attributed to the following: [22]

- Business Users: 150,000 businesses sending payments over the RTP network, a 50% increase since December 2022;

- Consumers: More than 3 million consumers sending A2A payments and Zelle® payments over the RTP network;

As of September 2023, the RTP network reached 65% of US Demand Deposit Accounts (DDAs). [23]

As the RTP network matures, TCH continues to roll out new functionality to increase the capabilities for financial institutions and their customers. Since its launch, enhancements to the RTP network have included: [23]

- Request for Payment (RfP) which allows financial institution customers to request a payment from another account holder at a participating institution;

- Document Exchange, enhancing the RTP network's capabilities by providing easy access to documents in the same transaction flow with the payment or RfP;

- Enhanced Account Number Security that supports tokenization of account numbers used to request or make payments; and

- An increase in the value limit for payments in the RTP network to $1 million. [23]

FedNow

In 2019, after an extensive assessment encompassing broad stakeholder input,, the Federal Reserve announced that it would build an instant payment system for the United States. The decision concluded an analysis that was undertaken in response to a Faster Payments Task Force recommendation that the Fed consider an operational role, driven by a widely held view that it would be necessary to achieve ubiquitous reach.

In this announcement, the Fed described how and why the planned FedNow instant payments service could co-exist with TCH's RTP, similar to the competitive multi-operator environment for ACH and wire transfer services.

The 2019 Federal Register Notice indicated that a single provider of RTGS services for retail faster payments without competition may create undesirable outcomes for pricing, innovation, service quality, and reach. Conversely, provision of the FedNow Service alongside private-sector RTGS service would give banks the option of choosing a service or connecting to more than one service, a choice they have today for all existing payment services. The presence of multiple RTGS services for faster payments could yield efficiency benefits such as lower prices, higher service quality, and increased innovation [25]

Consistent with its collaboration objective, the Fed established a FedNow Community and several work groups to help guide design and development of the service, addressing aspects such as::

- ISO 20022 message specifications

- Fraud prevention features

- Bill payment use cases

- Reconciliation and reporting needs

- Request for Payment requirements

- Requirements unique to:
  - Core Providers
  - Corporate Credit Unions
  - Banker's Banks
  - Brokerage firms

In October 2020, the Federal Reserve announced the creation of the FedNow Pilot Program to support development, testing and adoption of the FedNow Service, with more than 120 financial institutions and processors subsequently enrolling in the two and a half year program. [28]

The FedNow Service went live on July 20, 2023, with 35 participating financial institutions, 16 processors and the U.S. Department of the Treasury. By early December 2023, the network had grown to 331 participating financial institutions .

RTP and FedNow Interoperability

Instant payment clearing and settlement requires a multitude of systems to interact seamlessly, making common approaches to formats and processing critical for success. Indeed, interoperability has been a key focus of collaborative industry efforts throughout the instant payments journey in the United States. In particular, stakeholders have urged the Federal Reserve and The Clearing House to prioritize efforts to ensure that the FedNow and RTP services are interoperable. Acknowledging this industry expectation, the Federal Reserve stated in its 2020 Federal Register Notice that it cannot accomplish interoperability for instant payments alone. The industry—banks, bank service providers, and service operators—must work towards this common goal, as it has in the past with other payment services. [27]

One model of interoperability used in card payments and wire transfers is likely to be most relevant to instant payments. It relies on the sending bank routing payments to either RTP or FedNow based on a table of bank routing numbers supported on each. If the desired endpoint can be reached on either network, the sending financial institution can choose where to route the payment using criteria such as cost, network functionality or other factors. This model is already being deployed by service providers connected to both networks with minimal, if any, disruption to the sending or receiving financial institution.

A second, less likely, model requires cross-network clearing and settlement where financial institution participants can choose to just use either RTP or FedNow, with the network routing out-of-network payments to the other network operator. ACH uses this model involving bilateral exchange between TCH's EPN and the Federal Reserve's FedACH Service.

Conclusion

Roughly 10 years into the instant payments journey in the United States, the payments landscape has changed significantly, as have user expectations. With two instant payments infrastructures in place, the industry is poised for a new era of innovation and competition. Adoption of faster payment solutions, especially instant payments are accelerating quickly in the US, even with our late entry compared to other countries. The opportunity and challenge before the industry is to grow instant payment network reach and leverage the unique attributes afforded by instant payments to deliver innovative new services to businesses and consumers.

Although 2023 instant payment volumes are just a tiny fraction of total electronic payments in the United States, it is safe to say that the adoption curve for instant payments will be steeper than historical norms and faster payment solutions of all types will continue to grow and evolve into the future.

Chapter 3 - Understanding ISO 20022: The Universal Language of Financial Messaging

By Steve Wasserman

Introduction

In the ever-evolving landscape of international finance, communication is key. Financial institutions, banks, and businesses across the globe rely on a standardized language to exchange information efficiently and accurately. ISO 20022, often referred to as the "universal language of financial messaging," plays a pivotal role in ensuring seamless communication and data exchange in the modern financial world.

In this chapter we look at…

- What it is

- It's history

- Where it's used

- Why it's important

- What it consists of

- It's differences

- It's future

What it is

ISO 20022 is an international standard for financial messaging that defines a common platform for the development of messages in the field of financial services. It provides a standardized framework for creating and exchanging structured financial data electronically, fostering interoperability among various systems and institutions.[1]

Many financial institutions are in the process of migrating to ISO 20022 from older messaging standards. The migration requires careful planning, as it involves changes to systems, processes, and staff training. However, the benefits in terms of efficiency, accuracy, and future-readiness are significant.

It's history

ISO 20022 has a relatively recent history in terms of international adoption. Here's a brief overview of its history use internationally:

Early 2000s: Inception of ISO 20022

- The development of ISO 20022 began in the early 2000s with the aim of creating a modern, standardized messaging format for the global financial industry.

2004: Initial Publication

- ISO 20022 was first published in 2004, marking the release of the first edition of the standard. [2] This initial version laid the foundation for subsequent editions and updates.

Mid-2000s: Early Adopters

- Several European countries and financial institutions were among the early adopters of ISO 20022. The adoption in Europe was partly driven by the European Payments Council (EPC), which endorsed the standard for payments in the Single Euro Payments Area (SEPA). [3]

Late 2000s: Expansion and Global Interest

- ISO 20022 gained momentum as more countries and regions recognized its benefits for modernizing financial messaging. The standard's extensibility and flexibility made it appealing for a wide range of financial services beyond payments, including securities, trade finance, and more.

2010s: Widespread Adoption

- ISO 20022 saw significant growth in adoption across different regions and financial domains. It became a key component of initiatives such as SEPA, TARGET2, and many national real-time payment systems. [3]

2013: SWIFT Migration

- SWIFT, a global provider of secure financial messaging services, announced its plan to migrate its messaging platform and MT message standard to MX using ISO 20022. This move signaled the growing importance of ISO 20022 on the global stage. [3]

2017: Global Adoption Continues

- ISO 20022 continued to expand its presence globally, with more countries and payment systems adopting the standard. Many financial institutions embarked on migration projects to transition from older messaging standards to ISO 20022. 2020s:

Regulatory Mandates

Regulatory authorities in various countries, including the European Union, the United States, and others, began to mandate the use of ISO 20022 for specific types of financial reporting and payments, further cementing its status as an international standard.

Ongoing Evolution:

- ISO 20022 remains a dynamic and evolving standard, with regular updates and maintenance to meet the changing needs of the financial industry. Its adoption continues to shape the future of financial messaging and interoperability worldwide.

Today, ISO 20022 is recognized as the international standard for financial messaging, facilitating seamless communication and data exchange among financial institutions, businesses, and market infrastructures across the globe. Its widespread adoption reflects its importance in modernizing and streamlining financial services on a global scale.

Where it's used

ISO 20022 is in various stages of adoption and implementation across many countries and regions. The adoption of ISO 20022 can vary depending on the specific financial systems, payment networks, and use cases within each country. As of 2023, here are some of the countries and regions where ISO 20022 was being actively used or adopted:

- **European Union (EU):** ISO 20022 has been widely adopted in the EU, particularly for payments through the Single Euro Payments Area (SEPA) and the Euro system's TARGET2 platform.[3]

- **United Kingdom:** ISO 20022 is used in the UK for various payment systems, including the Faster Payments Service (FPS) and CHAPS[3]

- **Switzerland:** ISO 20022 is used for the Swiss Interbank Clearing System (SIC)[4] and the Swiss Euro Clearing System (SECOM)[5].

- **Nordic Countries:** ISO 20022 has been adopted for payments and securities across the Nordic countries, including Denmark, Sweden, Norway, and Finland.[6]

- **Canada:** Canada adopted ISO 20022 for its high-value payment system, the Large Value Transfer System (LVTS), and for the Lynx real-time payments system.[7]

- **Australia:** ISO 20022 is used for various payment systems in Australia, including the New Payments Platform (NPP) for real-time payments.[8]

- **Singapore:** ISO 20022 is used for Singapore's real-time payment system, FAST (Fast and Secure Transfers).[9]

- **Hong Kong:** Hong Kong has adopted ISO 20022 for its real-time payment system, the Faster Payment System (FPS).[10]

- **Japan:** ISO 20022 has been adopted for various financial messaging purposes in Japan, including securities and cross-border payments.[11]

- **South Africa:** ISO 20022 is used for the South African multiple message systems for interbank payments.[12]

- **Brazil:** ISO 20022 is used for various financial services, including securities and payment systems (PIX) in Brazil.[13]

- **United States:** ISO 20022 adoption has been growing in the United States. The Clearing House's (TCH) RTP® platform went live in 2017. The Federal Reserve's FedNow Service went live in July 2023. The migration of TCH's Chips and the Fed's FedWire are in the works.

As more widespread implementation of ISO 20022 continues to spread, it further realizes the benefits of ultimate global interoperability and all of the benefits that the implementation can mean to the end consumers and businesses that adopt solutions provided by the financial institutions and payment services that build products and use cases using it.

Why it's important

ISO 20022 is important for several reasons, as it offers significant benefits to the financial industry, businesses, and the global economy. Some key reasons why ISO 20022 is considered crucial include the following:[14]

- **Standardization and Global Adoption:** ISO 20022 provides a single, globally recognized standard embraced by financial institutions and market infrastructures worldwide, making it a truly global standard for financial messaging fostering consistency and interoperability in international financial messaging. It helps financial institutions operate seamlessly across borders and currencies.

- **Rich Data Structure and Data Quality:** ISO 20022 messages are designed to accommodate a wide range of financial transactions and carry comprehensive data, allowing for more detailed and precise information exchange. It's messages have a structured and extensible format, allowing for the inclusion of detailed and context-rich information. This data richness improves the accuracy and completeness of transactions, reducing errors and exceptions. It establishes a common language for financial institutions, businesses, and market infrastructures, reducing the complexity of communication and increasing interoperability. Through its dictionary of common terms for communicating the messages, the standard promotes data accuracy, consistency, and ultimately quality.

- **Regulatory Compliance:** Many regulatory authorities are mandating or endorsing the use of ISO 20022 for reporting and compliance purposes. The use of ISO 20022 rich data standards can help organizations meet regulatory requirements more efficiently.

- **Improved Analytics:** The structured rich data in ISO 20022 messages facilitates advanced AI powered analytics and data-driven decision-making. Financial institutions can gain valuable insights from transaction data as well as aid in detecting and mitigating fraud.

- **Enhanced Customer Experiences:** ISO 20022 enables businesses and financial institutions to provide more transparent and informative messages to their customers, improving customer service and satisfaction.

- **End to End Communications and Automation:** ISO 20022 facilitates end to end features and benefits, such as front end payment initiation starting from a Request For Payment (RfP) to back end reporting and reconciliation with accounts receivable remittance detail Straight Through Processing (STP).

- **Reduced Costs:** While initial implementation may require an investment, ISO 20022 can ultimately reduce operational costs through increased automation, streamlined processes, and reduced error handling.

- **Future-Proofing:** ISO 20022 is a forward-looking standard that can adapt to emerging technologies and changing industry needs. Organizations that adopt ISO 20022 are better positioned to address future challenges and opportunities.

In summary, the use of ISO 20022 is important because it modernizes and standardizes financial messaging, enhances data quality and accuracy, improves automation and efficiency, and supports the growth and innovation of the global financial industry. It is a key enabler for a more interconnected and digital financial ecosystem.

What it consists of

ISO 20022 consists of a set of standardized messaging standards and data definitions that are used for electronic communication in the financial industry. It defines the structure and content of financial messages, enabling consistent and standardized data exchange between financial institutions, businesses, and other entities involved in financial transactions. Here's what ISO 20022 consists of:[15]

- **Message Domain Types:** ISO 20022 includes a wide range of message types that cover various financial processes and transactions. These messages are categorized into domains, such as payments, securities, trade finance, and more. We are just focusing on the payment domain used for instant payments US market implementations.

- **Value and Non-Value Message Types:** Commonly used message types can be summarized as value and non-value messages.

  - **Value messages:** These involve the actual movement of funds.

    - **Credit transfers:** This covers customer credit transfers of funds from a customer account in one financial institution to a customer account in another financial institution. These can be initiated by a payer instructing their financial institution to send the funds with or without that payer acting in response to a request for payment initiated by the payee's financial institution. Credit transfers can also be between two financial institution's house accounts when it is not specific to a customer account on either end.

→ **Note:** The US instant payment rails only implemented credit push transfer transactions. Some international market implementations of ISO 20022 also enable real-time direct debit pull transactions. The implementation of just supporting credit push eliminates the fraudulent scenario of unauthorized debits. The ISO 20022 instant payment request for payment message is an alternative to the debit pull approach which provides more control and security for the payers and their financial institutions and payment services that no longer have to deal with disputes regarding unauthorized debits, though they may still deal with authorized push payment fraud scenarios and errors in the credit push transactions.

➤ **Returns:** This covers the return transfer of funds when the original credit transfer was sent in error (e.g. to an invalid destination customer account) or in response to a request for a return of funds that was sent as a result of being identified as a result of a fraudulent transaction.

⊙ **Non-value messages:** These are messages which do not transfer funds. Instead they either initiate value messages, respond to request messages, and cover other non-value message functions.

➤ **Request for Payment (RfP):** This is where a customer from one financial institution can request a credit transfer from a customer of another financial institution to pay them for the specifics of what is described in the request. The obvious use of RfP messages is to pay for a bill. RfPs can be used for other purposes as well, such as a person to person requests, merchants requesting payment for a purchase at the point of sale or transacted online, or as a request for a refund from a biller or merchant.

➤ **Response to Request for Payment:** This is the message type used to respond to an RfP. It can either acknowledge that it will be making the payment or otherwise reject it indicating a rejection code and optional description to send back to the requestor.

Acknowledgement of the payment request does not transfer the funds. It just acknowledges receipt of the RfP and the intent to pay it. It can enable initiation of a credit transfer to be either sent immediately or scheduled to be sent based on a due date and time indicated within the request for payment message. The payment can also be paid outside of the instant payment rails where the payer can reject the request indicating that it was or will be paid some other way.

➤ **Request for Information (RfI):** This message type has many uses, but commonly covers requests for additional information about credit transfers that were received, such as for the explanation of the purchase or bill/invoice remittance information that the payment was sent for. RfIs may also be common in conjunction with inquiries or disputes about RfPs.

➤ **Response to Request for Information:** This is the message type used to respond to the RfIs.

➤ **Account Inquiry and Reporting:** These messages cover various cash management inquiries and reporting needs.

➤ **Acknowledgement:** There are a number of different types and uses of acknowledgement messages, such as those acknowledging receipt of payment forwarded to the payee as well as those sent to the payer. Acceptance or rejection of RfPs are another type of acknowledgement message exchange.

These acknowledge messages are intended to be communicated back to the customers of the sending and receiving financial institutions.

> **Message Status Request & Response to Message Status:** These messages are between the network rail and the participating financial institutions or their third party service providers that connect them to the network. They serve as a confirmation or rejection of a message sent or received sent from the network or from the network participant back to the network. These messages also provide a means of requesting and responding to requests about the status of other messages sent and received through the network.

> **System and Administrative Messages:** These messages include network message broadcasts, network participant information, network connection status checks, and participant maintenance window notifications.

> **Remittance Advice:** This is the information that explains what a payment was paying for such as bill/invoice numbers and/or additional information. This non-value message type can be sent separately from the payment, and is not limited to payments sent through the instant payment rails. The same information that this standalone message type includes can also be embedded into a credit transfer and other value messages as well as through some of the administrative statement and activity reporting messages as well as through the response to request for information message.

• **Message classes:** ISO 20022 classifies the instant payment value and non-value message functions under the following commonly used classifications:

- ◉ **Payment Initiation (PAIN):** This class includes credit transfer initiation and request for payment functions.

- ◉ **Payment Clearing and Settlement (PACS):** This class covers customer and financial institution credit transfers, return of fund transfers, and payment message status.

- ◉ **Cash Management (CAMT):** This class includes account activity and balance inquiries and reporting, debit/credit posting notifications, information requests, message cancellation, return of funds requests, and message rejection notifications.

- ◉ **Administrative (ADMI):** This class covers network message broadcasts, network participant information, network connection status check, and participant maintenance window notifications.

- ◉ **Remittance Advice (REMT):** This has the standalone remittance advice message.

- **Message Structure:** Each ISO 20022 message type has a well-defined structure that specifies the format and organization of the message. The structure includes elements and data fields that carry information related to the transaction. These elements are defined using XML (Extensible Markup Language) syntax, making the messages machine-readable and easily processed by computer systems.

- **Data Fields and groupings:** ISO 20022 messages consist of various data fields, each with a specific defined purpose, usage, and format. These data fields are used to convey common groupings of information throughout all of the messages such as sender and receiver identification and address details, financial institution and account details, transaction amounts, transaction references, dates, and additional transaction specific information. The standardized dictionary of data fields and groupings ensures consistent and accurate data exchange.

- **Business Rules and Validation:** ISO 20022 includes a set of business rules and validation checks that must be adhered to when creating and processing messages. These rules ensure data accuracy and consistency and help prevent errors in financial transactions.

  - ◉ **Note:** Rail specific implementations of the standard can extend the business rules, vary specific message usage rules, mandate optional fields in the standard, and additional field level validations criteria. When implementing instant payments, you need to follow both the ISO 20022 standards as well as any rail specific implementation differences. Even though this complicates interoperability, its still easier to enable interoperability since all implementations are based on the common base standard.

- **Code Lists:** The standard includes predefined code lists that specify valid values for certain data fields. For instance, currency codes, country codes, and payment method codes are standardized to ensure consistency in message content. Some code lists have been externalized from those that are fixed and are part of the standard, which would require the standard being modified through one of its maintenance approval and update cycles. The process to add codes to the external codes lists enables more flexibility and timeliness for updates to these lists.

- **Metadata and Documentation:** ISO 20022 provides comprehensive documentation and metadata for each message type, including data definitions, field descriptions, and usage guidelines. This documentation is essential for organizations implementing ISO 20022 to understand how to create, interpret, and process messages correctly. Swift has a My Standards portal (https://www.swift.com/standards/mystandards-and-swift-translator) for financial institutions and others that need access to this ISO metadata and documentation

- **Maintenance and Updates:** ISO 20022 is a dynamic standard that evolves over time to meet the changing needs of the financial industry. It is subject to regular updates and maintenance by the ISO working groups responsible for its development. Updates and enhancements to ISO 20022, like any international standard, typically follow a structured process overseen by the ISO organization. The process involves various stages and timelines, and it aims to ensure that the standard remains relevant and responsive to the evolving needs of the financial industry. Here is an overview of the process and timing for updates and enhancements to ISO 20022:

- **Identification of Needs:** The process begins with the identification of needs for updates or enhancements. This can be driven by technological advancements, changes in industry requirements, regulatory developments, or feedback from users.

  - ⊚ **Proposal and Initiation:** A proposal for updating or enhancing ISO 20022 is submitted to the relevant ISO Technical Committee (TC) or Working Group (WG). The proposal includes details about the proposed changes, their rationale, and its expected impact.

  - ⊚ **Drafting and Development:** Once a proposal is accepted, a working group is formed to draft the updated or enhanced version of ISO 20022. This stage involves extensive discussions, reviews, and contributions from industry experts and stakeholders.

  - ⊚ **Public Consultation:** Draft versions of the proposed changes are made available for public consultation. Interested parties, including financial institutions, businesses, and regulatory authorities have the opportunity to provide feedback and suggestions.

  - ⊚ **Revision and Refinement:** The working group reviews the feedback received during the public consultation and incorporates relevant changes and improvements into the draft. This iterative process continues until a consensus is reached.

⊚ **Committee Draft (CD):** The revised draft is then circulated within the ISO TC or WG for further review and approval.

⊚ **Draft International Standard (DIS):** If the CD is approved, it moves to the Draft International Standard (DIS) stage. At this point, it is distributed to ISO member bodies for a vote on its approval.

⊚ **Final Draft International Standard (FDIS):** If the DIS is approved, it may go through additional revisions and refinement based on comments received during the voting process. The final version is then circulated as a Final Draft International Standard (FDIS) for approval.

⊚ **Publication:** Once the FDIS is approved, the updated or enhanced version of ISO 20022 is published.

⊚ **Implementation:** After publication, organizations, financial institutions, and payment systems may begin implementing the updated or enhanced ISO 20022 standard as part of their systems and processes.

The timing for updates and enhancements to ISO 20022 can vary significantly depending on the complexity of the changes and the consensus-building process. It can take several years from the initial proposal to the publication of a new version.

It's differences

You may wonder what's different about ISO 20022 from other payment standards, such as SWIFT MT, NACHA, and ISO 8583.

Differences compared to **SWIFT MT**'s legacy format include: [16]

- **Structure and Data Richness:** ISO 20022 messages are structured in Extensible Markup Language (XML), a versatile and flexible data format. This allows for highly structured and rich data, supporting a wide range of financial transactions with extensive details. ISO 20022 messages can accommodate complex data requirements, making them suitable for modern financial services, including instant payments and regulatory reporting. SWIFT MT messages, while widely used in the past, are less structured and have limited space for additional data. They are less adaptable to the growing complexity of financial transactions and often require supplementary documentation.

- **International Standard:** SWIFT MT messages are specific to the SWIFT network and were primarily designed for international communication. While they have been widely used in the USA for cross-border payments, they are less standardized globally compared to ISO 20022 where it is being implemented for domestic as well as cross-border payment networks. Even SWIFT has started to convert from its MT format to MX, which is its implementation of the ISO 20022 standard.

- **Versatility and Extensibility:** ISO 20022 is highly versatile and extensible, allowing implementing networks to customize message definitions to meet specific local, regulatory, and business requirements. SWIFT MT messages are less extensible and require additional codes or workarounds to accommodate these types of variations. Customization options are limited compared to ISO 20022.

- **Support for Modern Financial Services:** ISO 20022 is well-suited for modern financial services, including real-time or instant payments, regulatory reporting, and advanced analytics. It can accommodate emerging technologies and evolving industry needs. SWIFT MT messages may lack the data richness and structure needed for modern services like instant payments, which require real-time data exchange and enhanced information sharing.

- **Regulation and Compliance:** ISO 20022 is increasingly becoming a regulatory requirement in the financial industry, including in the USA. Regulators see it as a means to improve transparency, reduce risk, and enhance reporting capabilities. While SWIFT MT messages have been used in compliance and reporting, they may require additional effort and reconciliation due to their limited and relatively fixed data structure.

Differences compared to the **NACHA** and **ISO 8583** standards include: [17]

- **Scope and International Adoption:** ISO 20022 is a versatile and comprehensive international standard for financial messaging. It covers a wide range of financial services, including payments, securities, trade finance, and more. ISO 20022 is widely adopted globally, and it promotes interoperability by providing a common language for financial messaging across different countries and regions.

    NACHA, which stands for the National Automated Clearing House Association, is a U.S.-specific organization and rulebook governing the Automated Clearing House (ACH) network. NACHA rules primarily govern domestic payments in the United States. It has been adopted in a few other countries, but it is not considered a standard that has been embraced internationally as ISO 20022 has. NACHA did add an international message class to its specification (called IAT), but it has had very limited adoption and it is only used for multicurrency transactions settled through US bank accounts.

    ISO 8583 is a messaging standard primarily used in the card payment industry, particularly for ATM and retail point of sale (POS) and e-commerce transactions. It focuses on card-based payments and authorization requests. ISO 8583 is widely used in the card payment industry globally and has been adapted for various regional payment card systems.

- **Structure and Data Richness:** ISO 20022 messages are structured in Extensible Markup Language (XML) and offer a highly flexible and rich data format. This structure allows for detailed information and supports complex financial transactions making them suitable for modern financial services, including real-time payments, regulatory reporting, and advanced analytics.

  NACHA messages are often transacted using relatively limited fixed file formats which are less rich and versatile than ISO 20022 messages. There are a few uses of the NACHA format using XML, but these still are limited to the limited data fields that the NACHA standard includes.

  ISO 8583 messages have a fixed-length format with predefined fields. This format is designed for efficiency in high-volume card transactions and is less flexible compared to ISO 20022.

It's future

Let's take a look into the future of ISO 20022, **at least at the point of time when this book was written during the summer into the early fall of 2023.**

The future includes in-progress and planned implementations of the ISO 20022 standard for the US domestic Fedwire and TCH's CHIPS wire services as well as for SWIFT for international cross-border payments. The timeline for the implementation of ISO 20022 in networks varies by network, region, and project scope. Here are some key timelines and milestones for these implementations:

- **SWIFT's Migration to ISO 20022 for Cross-Border Payments:** This was announced in 2013 with the following implementation migration plan from 2016 to 2025.[18]

  ◉ 2016-2019: SWIFT launched a pilot phase and began introducing ISO 20022 standards for specific market infrastructures and correspondent banks.

- 2022: SWIFT planned to introduce ISO 20022 for cross-border payments on a phased basis, starting with certain corridors.

- 2023-2025: SWIFT planned to complete the migration across all correspondent banks and corridors.

- **Implementation in Domestic Wire Transfer Systems:** This varies by country which may have its own schedule and approach for adopting ISO 20022. Many countries opt for a phased approach, migrating specific payment systems or payment types to ISO 20022 over time. This allows for gradual adoption and minimizes disruptions. In some cases, regulatory authorities have or may mandate the adoption of ISO 20022 for specific payment types or reporting requirements, which can accelerate the implementation timeline. Implementing ISO 20022 in domestic wire transfer systems often involves extensive planning, coordination with financial institutions, and testing phases. ISO 20022 implementation projects in domestic wire transfer systems may continue to evolve as new payment types and services are introduced.

- **U.S. Fedwire:** The Federal Reserve initiated a project to migrate its Fedwire Funds Service to ISO 20022 messaging format. The migration was expected to occur in multiple phases, with the initial phase targeted for 2023 and completion in 2025. The adoption of ISO 20022 in Fedwire was seen as a significant step in modernizing the U.S. payments infrastructure and enhancing interoperability with other payment systems.[19]

- **TCH CHIPS (Clearing House Interbank Payments System):** The Clearing House announced plans to migrate CHIPS to ISO 20022 messaging standards by April 2024. The migration project aimed to align CHIPS with global payment messaging standards, improve the efficiency of cross-border transactions, and enhance the overall capabilities of the payment network.[20]

Next we will look at what future types of ISO 20022 implementations are in the works or planned. The adoption of ISO 20022 was and continues to be driven by the need for modernization, increased data richness, and improved interoperability. Some trends and areas where ISO 20022 is expected to play a significant role in include:

- **Real-Time Payments:** Many countries were either in the process of implementing or planning to implement ISO 20022 for their real-time payment systems.

- **Cross-Border Payments:** ISO 20022 was seen as a way to improve the efficiency and transparency of cross-border payments.

- **Open Banking:** ISO 20022 can facilitate data exchange and interoperability in open banking ecosystems, allowing financial institutions to share customer data securely and efficiently.

- **Corporate-to-Bank and Bank-to-Corporate Communication:** Corporates and businesses are increasingly starting to use ISO 20022 for their interactions with banks and financial institutions, improving the efficiency of treasury and cash management processes. Use of the standard also enables more standardized flexibility to interface their systems with multiple or new banks domestically as well as internationally.

- **End to End Corporate-to-Corporate Communications:** Some of the message types, such as the Request for Payment and Standalone Remittance messages also lend themselves to B2B trading partner direct or indirect communications. The indirect route would be to and from their financial institution or payment services on both ends of the message transmission and receipt. The more that businesses can send and accept the standard messages with their trading partners, the more benefits, including potentially lower costs, can be realized up and down their supply chains. The adoption of ISO 20022 in trade finance could also improve the efficiency and transparency of trade-related messaging and documentation.

- **Central Bank and Other Digital Currencies (CBDCs, Stablecoins, Crypto):** As central banks explore the issuance of CBDCs, ISO 20022 could play a role in the messaging and settlement infrastructure for these digital currencies. Some stablecoin and even crypto platforms already include or plan to incorporate the use of ISO 20022 to enable some level of interoperability with other payment systems, if not otherwise just used to on and off ramp to these digital currency platforms.

The future for ISO 20022 appears promising and dynamic, with ongoing developments and increasing adoption across the financial industry. Here are some key aspects that define the future success of the value for ISO 20022:

- **Widespread Adoption:** ISO 20022 adoption continues to expand globally across various financial domains, including payments, securities, trade finance, and more. As more countries and regions embrace ISO 20022, it is becoming a de facto standard for financial messaging.

- **Regulatory Mandates:** Regulatory authorities in different countries are mandating the use of ISO 20022 for various reporting and compliance purposes. Compliance with ISO 20022 standards is expected to become a regulatory requirement in more jurisdictions. In countries where these mandates have been made, they have seen faster and more widespread adoption than where not otherwise mandated. These countries have thus benefited from this more than those that have not set mandates. In other countries, industry level groups that have collaborated to adopt ISO 20022, have seen more gradual adoption and some only act on a basis of FOMO (fear of missing out).

- **Enhanced Data Analytics:** The structured and rich data content of ISO 20022 messages enables advanced data analytics and reporting capabilities. Financial institutions are using this data to gain valuable insights and improve decision-making.

- **Digital Transformation:** ISO 20022 is a key enabler for digital transformation in the financial industry. It supports the adoption of emerging technologies such as blockchain, artificial intelligence, and machine learning. These digital transformations are also happening in businesses adopting the standard or otherwise leveraging the benefits of payments transacted through ISO 20022 that help with automate and streamline their banking and payment as well as trading partner digital information exchanges.

- **Globalization:** ISO 20022 promotes global interoperability by providing a common language for financial transaction messaging. This enables smaller and regional financial institutions to participate more effectively in global markets. This also significantly enables businesses of all sizes to transact with trading partners around the world.

- **Cybersecurity and Fraud Prevention:** The rich data in the ISO 20022 messages can be leveraged to enhance cybersecurity and fraud prevention efforts through extended and improved data quality and transparency.

- **Interoperability:** Efforts are ongoing to improve interoperability between different payment systems and networks that use ISO 20022, making cross-network transactions more seamless. Similarly, the use of the standard by corporates enables more interoperability with various types and sizes of trading partners.

- **Future Revisions:** ISO 20022's flexibility and continued enhancement capabilities enable it to keep up with emerging industry requirements, technologies, and regulatory changes.

Conclusion

ISO 20022 is a cornerstone of modern financial communication, offering a standardized and versatile framework for exchanging financial data globally. Its adoption continues to grow, enabling financial institutions to meet the evolving needs of a digital, interconnected world. As the financial landscape continues to evolve, ISO 20022 will remain a critical tool for ensuring seamless and efficient financial messaging.

ISO 20022 is a standardized framework for financial messaging that consists of message types, message structures, data fields, business processes, and validation rules. It ensures the consistent and efficient exchange of financial data across various domains in the global financial industry, promoting interoperability and automation while reducing errors in financial transactions and enabling new uses of payment related messages across end-to-end transaction communications.

Instant payment systems using ISO 20022 messaging typically encompass both value and non-value message classes. These message classes serve various purposes within the instant payment ecosystem.

ISO 20022 is a modern and globally recognized standard that offers greater data richness, versatility, and extensibility compared to legacy messaging standards like SWIFT MT, NACHA, and ISO 20022. None of these have the same rich data, versatile, and flexible characteristics. They all primarily use limited fixed file formats versus ISO 20022's use of XML's expandable and flexible format.

The scope of SWIFT MT exclusively for cross-border payments is limited compared to ISO 20022 which can be used domestically as well as for cross-border transactions. NACHA is specific to the United States (plus supported in a couple of other countries) and primarily governs domestic ACH payments within the U.S. ISO 8583 is designed for card-based payment transactions and is just widely used in the card payment industry worldwide.

The choice of which standard to use depends on the specific requirements of the financial transactions and the region or industry in which they are applied.

The future for ISO 20022 is characterized by its increasing importance as the global standard for financial messaging. Its versatility, rich data structure, and adaptability to modern financial services are driving its continued adoption and evolution. As the financial industry embraces digital transformation and strives for greater interoperability and transparency, ISO 20022 will play a central role in shaping the future of financial services worldwide.

As implementations of ISO 20022 continue to spread, it further realizes the benefits of ultimate global interoperability. Consumers and businesses also benefit from products provided by the financial institutions and payment services that incorporate the functionality and data communicated through ISO 20022 messages.

It's important to note that the timelines for ISO 20022 implementation can vary based on regional factors, industry initiatives, regulatory requirements, and the complexity of existing infrastructure. Organizations involved in these implementations typically work closely with industry bodies, central banks, financial institutions, and payment service providers to ensure a smooth transition to ISO 20022 messaging standards. To get the most up-to-date information on specific ISO 20022 implementation timelines, it's advisable to consult with relevant authorities or organizations in your region or the regions of interest.

ISO 20022 is an evolving standard, and updates may occur on a periodic basis to ensure that it remains relevant and aligned with industry needs and technological advancements. Organizations that rely on ISO 20022 should monitor updates and consider how they will impact their operations and systems.

It's also important to understand the adoption of ISO 20022 varies by region, network, financial institution, payment service providers, and corporate businesses. New implementations and use cases continue to emerge as the standard gains traction worldwide. To stay updated on specific future market implementations and developments, it's advisable to follow announcements from central banks, financial market infrastructures, regulatory bodies, financial institutions, and industry organizations in your region or the regions of interest.

With instant payments adoption in the US, The strengths and benefits of ISO 20022 are just icing to the cake of all of the benefits and functionalities that instant payments already provides in terms of speed and their 24x7x365 availability.

Chapter 4 - Implementations: Strategic Approach & Go To Market Plan[1]

By Travis Dulaney

Building the Go To Market Plan

This chapter is written to help Fintechs navigate the complex and challenging area where technology and regulation meet and how to think about building your banking relationships for a successful instant payment implementation program and utilize various forms of real-time payments or embedded finance functionality into their product offerings.

A critical component to the success of your implementation strategy is contingent on aligning your approach with your intended objectives and having a good understanding of your overall "use cases" and the stakeholders.

But first take time to outline your approach. The following areas are critical to understand when building a relationship with a bank and or service provider.

1. **Time to Market:** The efficient launch of your program relies on execution by all parties that are part of your use case and

business model constructs. Your time to market could have a big impact on who you partner with to build out your program.

2. **Price Point & Market Pricing:** While you don't need to have all your pricing finalized before getting started, you need an idea of what you plan to charge for the service or how you plan to monetize at a minimum.

3. **Program Scalability:** Preparing for expansion to meet growing demand is essential. So how you assemble your partners and third party provider's matters if you are looking at high volumes of transactions.

4. **Compliance Readiness:** Compliance cannot be an afterthought or otherwise you will never get past the first gate with a bank or any creditable service provider. This is a regulated process and coming to the table with the appreciation of what is required to be compliant is a must!

5. **Technology Limitations:** Grasping potential technological constraints and opportunities is key. Not everyone has unlimited budgets or resources, knowing what you can and can't accomplish will save you pain later with false starts of your program.

6. **Vendor Management & Third-Party Providers:** A quick way to get to market faster sometimes is through partners and third party providers. Establishing robust relationships enhances program reliability but also increases your operational responsibilities related to vendor management and regulatory compliance oversight.

7. **Product Maturity:** Depending on the maturity of your product, you could design the service offering differently. This could mean working with different partners and most importantly adjusting current internal processes and workflows, and user payment experiences.

For example, a corporation like an Insurance company that is currently using customer ACH routing & account numbers as part of their payment process for paying insurance premiums may elect to use the same information for claim disbursements so they do not ask the customer for the same information twice.

Establishing clear expectations regarding the above facets is a fundamental prerequisite to getting started. Neglecting this vital step can result in wasting time and resources. The journey from program inception to market launch typically spans a minimum of six months, often extending to over one year depending on the complexity and risk associated with your program.

When you first set out to consider implementing real-time payments, it is helpful to understand that there are multiple ways to "plug into the ecosystem" and how you connect to the ecosystem will further define your time to market, cost and operational requirements.

The diagram below provides you a simple look at the payment industry supply chain for instant payments.

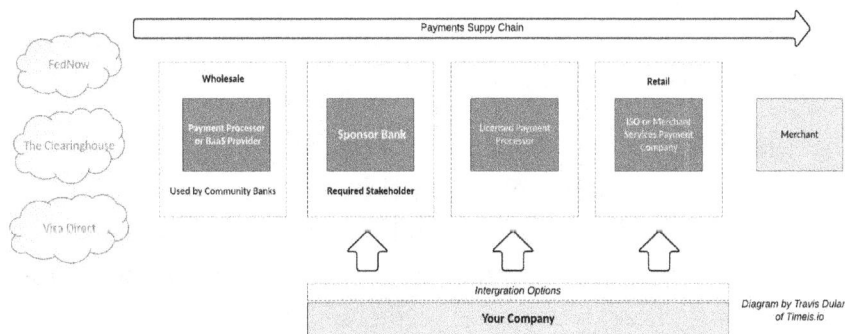

Generally speaking, connecting closer to the networks directly with a bank will lower the cost of your transactions but working with partners and payment providers will help you get to market faster since they have worked through the regulatory compliance and operational challenges already.

Another notable exception to a long engagement timeline is to build a relationship with a group of pre-approved processors, known as PayFacs. PayFacs typically process credit cards only but for the ones that have built a relationship with a Banking As A Service Sponsor Bank, they could be a valuable solution if you need to get to market quickly.

PayFacs have gotten their "use case" pre-approved with the bank, appropriate networks and organizations. While PayFacs expedite time-to-market, their utilization can have adverse effects on margin, market pricing, and long-term scalability. Choosing a growth path, whether as a startup or an established corporation, is a strategic decision that warrants thoughtful consideration.

As the age-old adage goes, "Measure twice and cut once." Diligently crafting your approach and go-to-market strategy will yield long term dividends.

The banking and payments space operates within a complex and tightly regulated environment. It's important to note that banks operate as risk management entities, grounded in calculated risk taking principles and risk management processes and procedures. So be prepared to be patient.

Facilitating payments and fund transfers may appear straightforward, but it is among the most inherently risky decisions that banks make. Stringent regulatory oversight underscores the need for caution and appreciation when interacting with banking institutions. Your awareness of these dynamics is pivotal in effectively engaging with your chosen bank partner or provider.

Characteristics of Faster Payments Program Implementation

When strategizing for the implementation of your real-time payments program in a business context, meticulous consideration, evaluation, and selection are essential across the following dimensions:

1. Partner Ecosystem

The process of selecting partners is a critical step that can significantly impact your operational efficiency and time-to-market. It requires a departure from merely recognizing well-known brand names. In fact, the most impactful partners may not be widely known; some of the best service providers are small growing startups that have unique and differentiating functionality that could give you a leg up with your project.

To make informed decisions, go beyond surface assessments and focus on building relationships founded on trust, accessibility, and reliable support. The true value of a partnership becomes evident when you can easily communicate with knowledgeable individuals, eliminating the frustration of email obscurity.

Your typical partner ecosystem comprises:

- Sponsor Bank
- Money Movement Processor
- Identity Validation Service
- Risk Management & Compliance Service
- Fraud Service Provider
- Third-Party Auditors
- Security Service Provider

2. Technology Framework

The dynamic nature of any technology platform requires a deep understanding of how your existing technology infrastructure aligns with the new components. The shift towards Instant Payments and Embedded Finance emphasizes the integration of various providers and processes throughout the customer lifecycle, while the conventional backdoor API approach allows for the partner development team to take liberties with the customer onboarding process. However, partner Fitness must follow Bank Security Act (BSA), Anti Money Laundering (AML) & Office of Foreign Assets Control (OFAC) regulatory guidelines along with other Information Security and Fraud protection best practices.

Embrace changes to your Security and Networking architectures to facilitate seamless data flow between your operations and your providers. Be aware of data definitions and data incompatibility issues and focus on transformation to well-defined data structures, helping to improve data integrity throughout the entire lifecycle of the financial transaction.

3. Legal Agreements

Once the Partner Ecosystem is defined, it transforms into a collection of agreements and contracts that drive the implementation process. This section underscores the importance of legal agreements with providers, extending beyond the scope of standard contracts.

Technically, from a regulatory perspective, your stakeholders in the process are not partners, but rather third-party service providers. Your agreements with the sponsor bank and payment providers all will have heavy language and provisions that require compliance with laws, regulatory rules and operational protocols, especially concerning data management and security, particularly end customer data. Keep in mind there are currently 26 laws that require your adherence too when your customer is an individual consumer.

So it goes without saying your customer agreements may need updating, including intertwining risk management and compliance requirements with your third-party provider agreements.

Avoid the misconception of evading financial risk and liability. Banks thrive on risk management, while processors have typically acted as fee collectors. In the current heightened regulatory environment, it is critical to understand you can't pass on your regulatory compliance responsibilities to another party, every entity is fully responsible for their own regulatory compliance and abiding to the laws based on the business model of your company. Understanding your risk spectrum and navigate it adeptly through robust controls and policy documentation, safeguarding your interests as you pivot towards revenue generation or cost reduction.

Incorporating a robust risk and compliance framework whereby your employees develop a culture of compliance becomes paramount in securing your financial operations and will keep your banking relationships strong. This strategic approach ensures that your real-time payments program implementation is not only efficient but also compliant and risk-aware.

Your Market, Customers, and Personas

Gaining a profound understanding of your target market extends beyond marketing initiatives; it plays a pivotal role on the journey of establishing a payment program or integrating financial services into your existing product or service offerings.

You might be wondering what does my product or service's target market and customer have to do with my real-time payments program?

It has everything to do with it! The type of customer and the industry that you operate your business in, can greatly impact how and who wants to support your business from a financial perspective.

How your company does business and with who matters. Just as we mentioned earlier, if you are a C2B company, there are over 26 laws that need to be considered when processing payments vs a B2B model.

Specific industries are considered higher risk than others, and some banks and providers elect not to support those industries due to the operational cost required to manage the risk.

High risk isn't a bad thing, but you do need to find the right partners to work with that best understand how to manage those risks. Finding partners in industries such as Digital Assets, Cannabis, Gambling, Money Service Bureaus (MSB), Pawn Shops are challenging at best but there are banks and payment providers that are focused on those industries.

Part of understanding your customer is understanding how the money will flow from one account to the next, depending on what you are trying to accomplish.

The Bank and/or Payment Provider will place a lot of emphasis on scrutinizing the origin of the funds effectively called the "source of funds". Due to Anti-Money Laundering (AML) regulations, you need to know whose money is being moved. As part of all payments transactions, validating that the funds being moved are not coming from any entity on one of the many governmental sanction lists is all part of the process so your partners will need to validate how and who you are doing business with, but most importantly, it is your responsibility, even when someone else is doing the validation, you are responsible and own the liability.

Precisely defining your customer's characteristics, geographic location, and demographics enables the creation of a comprehensive customer persona. This, in turn, provides the bank and payment partners clear insight into who your customer is but also empowers the detection of deviations throughout the customer's lifecycle, allowing you to clearly identify anomalies that could be fraudulent.

This helps serve as the foundation for identifying transactional patterns and historical behaviors to safeguard your company from fraud, malicious actors, and fostering a shared comprehension of how your sponsor bank and processor partners perceive your enterprise.

It is important to understand that as a business or fintech provider, looking to provide money movement and payment functionality to your product or service, most look to own control of the entire user experience so it is natural that you will want to gain more control of the customer onboarding process. Unfortunately, in our current regulatory environment you might not get that option of owning the customer experience since as a fintech or business providing payments as a service, you are basically considered a branch of the bank and will be regulated as if you were a bank.

The reason customer onboarding has so much compliance activities associated with it is that this is where the Know Your Customer (KYC) Anti-Money laundering (AML) & Office of Foreign Asset Control (OFAC) regulations come in. The Dodd–Frank Wall Street Reform and Consumer Protection Act, commonly referred to as Dodd–Frank, is a United States federal law that was enacted on July 21, 2010, outlining a process for Bank Security Act (BSA) which defined the controls required for onboarding banking account customers and as a extension of the bank, you must also comply.

In the last couple years, many banks allowed their third party providers to perform their own KYC processes. This did not meet the intention of regulations, so now, the regulators are coming down hard on the banks to manage the process better which has resulted many banks to only perform the validation checks themselves vs relying on the provider to be compliant along with taking on the responsibility of the bank being required to have effective third party vendor management process and procedures in place.

This validation process applies not only during the initial customer onboarding, but also when customers update their recorded information. The contemporary threat of account takeover looms large, making this rigorous validation crucial in safeguarding both your customers' accounts and their Personally Identifiable Information (PII) data.

A subsequent chapter provides comprehensive guidance on protecting your organization against risks and fraud. They illuminate effective strategies and measures for fortifying your enterprise in a business context.

Regulations and Fostering a Culture of Compliance

In the world of payments and banking, one is immediately confronted with a landscape of laws and government regulations. Within this complex framework, various government agencies hold distinct mandates to enforce, monitor, and occasionally impose penalties on organizations that fail to adhere to the prescribed regulatory guidelines.

The initial challenge lies in identifying the relevant regulations pertinent to your business. It's a nuanced task influenced by your market, business model, products and services offered and, critically, the flow of funds within your operations. Although this may seem straightforward, countless entrepreneurs and project managers have grappled with aligning their payment strategies with regulatory requirements.

In this book and this chapter, we will outline the diverse regulations relevant to payments and money movement. This journey commences with the establishment of a meticulously crafted Compliance Framework, tailored to align with your banking sponsor and the relevant laws and regulatory rules and guidance.

The foundational components of your Compliance Framework include:

1. **Regulation Alignment:** Precisely mapping relevant regulations to your unique operational landscape.

2. **Policies:** A comprehensive spectrum of policies encompassing KYC, AML, Sanctions (OFAC & other sanction lists), Information Security (InfoSec), Privacy, and Incident Management, among others.

3. **Work Process Controls:** Implementing controls that align operational practices with regulatory directives.

4. **Standard Operating Procedures:** A robust suite of procedures spanning training, financial, and operational processes.

5. **Repeatable Evidence Production:** Developing mechanisms to consistently produce evidence of compliance.

It is crucial to differentiate between security and compliance. While these two endeavors share overlaps, they remain distinct. A parallel can be drawn to the convergence of legal and compliance considerations within your end customer agreements and internal operational workflows.

The ultimate achievement in this complex endeavor is the construction of a compliance framework that does not burden employees or strain company resources. By adeptly constructing efficient processes and procedures, a significant portion of the operational workload can be automated. This approach leverages every team member's role as vigilant stewards, alert to anomalies and deviations from compliance norms, fostering a potent Culture of Compliance within the organization.

In essence, navigating and embracing the intricate web of regulations that best supports your end customers while protecting you partners and stakeholders in the process is a journey towards establishing a robust culture of compliance—a culture that not only safeguards against legal risks but also propels operational excellence, data integrity and ultimately customer satisfaction..

Navigating Bank Underwriting

First and foremost, in the world of payments and money movement, EVERYTHING comes down to a bank. By law, the only organizations that can move money across state lines is a bank or a licensed money transmitter.

So in preparation for engaging with a bank and seeking approval for your payment program you must start with a foundation of regulatory compliance. The importance of this stage cannot be overstated, as your endeavor hinges upon aligning with stringent regulations. The misconception that a compelling pitch alone can secure bank support is swiftly dispelled when dealing with a risk oriented banking environment.

Even if you work through a payment provider partner, they have a bank sponsor to answer too so you will still need to operate within applicable regulations and risk requirements of the partner's bank. In turn, any of your partners that are involved in transaction chains to your direct or indirect banks need operate within compliance and risk requirements

In essence, banks operate as bastions of risk management, overseen by professionals who prioritize regulatory adherence and fraud protection over entrepreneurial revenue enthusiasm. Thus, a bank's risk tolerance is meticulously calibrated to shield against adverse outcomes and within your attempt to be different or do things differently, you may be just causing yourself more challenges and conveying those differences and how they don't change the bank's optics on risk management you may find futile.

Straying from this understanding can lead to dire repercussions. Engaging with a bank requires an unswerving adherence to facts, an approach reminiscent of the classic line from the 1967 TV show Dragnet: *"Just the facts, ma'am"*. The objective isn't to peddle a sales pitch but to convey your business' customers, products and services and illustrate an intricate comprehension of your compliance strategy, and risk management efforts that manage the inherent risk in your business model thus addressing the bank's foremost concerns.

The following groundwork should be outlined before meeting a potential bank sponsor or provider.

1. **Business Case:** Present a succinct overview of your company's inception, location, target market, services, and essential business metrics.

2. **Business Model & Program Purpose:** Define the contextual purpose behind your payment system, value-added functionality, or fund movement. Emphasize the "what" rather than extolling the benefits—a perspective akin to reverse marketing.

   • Define your Products, Services and Customers along with the geographies that both the customer and the transaction is conducted within.

   • Define and communicate your Distribution Channels, Affiliates and any Stakeholder including value add partners that are helping you along the customer journey of your product and service. Think in two terms; (1) where the data flows and to whom. (2) Who is being compensated or benefits from the process. Obviously, Information Security (InfoSec) and protection of Personally Identifiable Information (PII) is a key element of understanding your program and its related risks.

   Many companies seem to think, if our business model isn't in the flow of funds, they are not subject to regulatory compliance but that is a wrong assumption, you might not have the same requirements as a payments processor but you still need to comply with regulations, rules and laws depending on what role you are playing in the process.

3. **Customer Onboarding:** Provide detailed insights into your validation procedures, mechanisms and fraud procedures to deter malicious actors. Have your controls clearly defined and outlined for how validate the person is who they say they are and what checks and validations you perform during the process of providing your customer an account. Be ready to provide evidence of your controls while highlighting both existing measures and planned enhancements, along with clear timelines for their implementation.

As with any business you have a contractual agreement with your customer to provide services. Within the world of payments and banking you will need written terms of service agreements and other customer disclosures based on your business model that define your legal terms with the customer. You will also need to define what compliance and data protection controls you have in place that protect your customers.

Depending on who your customer is and if they are an individual or business entity, be prepared to provide these documents to your providers.

4. **Compliance Framework:** Outline the regulatory landscape you adhere to, your policies and controls steering your operational processes that align with your business model, customers, products and services. Particularly emphasize the protocols entailing payment, banking, and customer data, including the management of the following:

Banking Regulatory Environment

Regulatory Rules & Consumer Protection Laws

Name	Reference
Know Your Customer	KYC
Bank Security Act	BSA
Personally Identifiable Information	PII
Anti-Money laundering	AML
Office of Foreign Asset Control	OFAC
Information Security	InfoSec
Payment Card Industry Security Standards	PCI
Service Organization Control 2	SOC 2
Statements on Standards for Attestation Engagements	SASE16
Network Rules	Card Network Rules NA-CHA Rules
Americans with Disability Act	ADA
Children's Online Privacy Protection Act	COPPA
Community Reinvestment Act	CRA
Consumer Leasing Act	N/A
Controlling the Assault of Non-Solicited Pornography and Marketing Act	CAN-SPAM
Electronic Funds Transfer Act	EFTA
Equal Credit Opportunity Act	ECOA
Expedited Funds Availability Act	EFAA

Fair Credit Reporting Act	FCRA
Fair Debt Collection Practices Act	FDCPA
Fair Housing Act	FHA
Flood Disaster Protection Act	FDPA
FTC Section 5 Unfair & Deceptive Acts or Practices	UDAP
Garnishment of Accounts Containing Federal Benefits Payments	N/A
Unlawful Internet Gambling Enforcement Act of 2006	UIGEA
Unlawful, Abusive, & Deceptive Acts or Practices	UDAAP
Uniform Commercial Code regulations	UCC
State Finance/Lending License Laws	N/A
State Money Transmission Licensing Laws	N/A
State Privacy Laws	N/A
Telephone Consumer Protection Act & Junk Fax Prevention Act	TCPA
Durbin (II)	N/A
Gramm-Leach-Bliley Act Privacy of Consumer Information	GLBA
Health Insurance Portability and Accountability Act	HIPAA
Home Mortgage Disclosure Act	HMDA
Military Lending Act	MLA
Real Estate Settlement Procedures Act	RESPA
Right to Financial Privacy Act	RFPA
Safe Act - Fraud/Identity Theft Red Flags	Safe Act
Secure & Fair Enforcement for Mortgage Licensing Act	Safe Act - Mortgage
Servicemembers Civil Relief Act	SCRA
Truth in Lending Act	TILA
Truth in Savings Act	TISA

With the growing trend towards cloud computing and outsourcing, SOC 2 reports are becoming a requested framework to comply with more over the recent years. SOC 2 compliance allows a service organization to provide assurance to its stakeholders that the service is being provided in a secure and reliable manner.

It's a security framework that defines how companies should manage, process, and store customer data based on the Trust Services Categories (TSC). There are five categories to adhere to;

- Security

- Availability

- Processing integrity

- Confidentiality

- Privacy

SOC 2 compliance is the closest we have to a general information security standard in the US.

5. **Corporate Website Readiness:** Mitigate the potential clash between marketing endeavors and compliance imperatives on your corporate website. A judicious balance is needed to avoid raising suspicion among risk managers who may peruse your website during their evaluation.

   Forward thinking statements mentioning products and services not currently offered should not be listed on your corporate website, it only introduces potential risk to your business model.

6. **Risk Management:** Cultivate an objective perspective on the inherent risks intrinsic to your business model and procedures. Engage in open, candid conversations about your risk mitigation strategies to underscore your commitment to

maintaining effective and repeatable controls as part of your operations..

7. **Financial Stability and Controls:** While financial stability is crucial, a robust balance sheet is not the sole determinant. Place equal emphasis on operational and financial controls. Learn from cautionary tales such as the Crypto company FTX, whose financial solidity was overshadowed by a lack of comprehensive controls.

In the realm of bank underwriting and due diligence, it is imperative to wield a clear comprehension of your compliance strategy as a formidable asset. By fostering a culture of adherence to rigorous regulations, you stand poised to establish a robust partnership with your chosen bank or provider, enabling the secure movement of funds in line with the highest standards of integrity and compliance.

Technical Integration

Aligning Technology and Purpose in Fintech

With a quarter-century of experience in the dynamic payments landscape, I often find amusement in witnessing the latest generation of tech developers placing an exclusive spotlight on technology. While undoubtedly a vital ingredient, the realm of Fintech underscores that the formula for success transcends pure technological prowess.

In the intricate tapestry of Fintech, technology indeed occupies a pivotal role, but the crux of the challenge is a delicate balance of customer delight, behavioral insights, satisfaction, and a clearly defined purpose. This symphony must resonate in harmonious alignment with the intricate fabric of banking laws, regulatory frameworks, and the bedrock of all relationships—the establishment of trust with your partners. Only once these foundational pillars are firmly in place does technology come to the forefront.

Navigating Integration Platform Considerations

The strategy for architecting the optimal platform is intrinsically tied to the purpose behind the integration. It's imperative to recognize that there exists a diverse array of pathways to integrate with different partners and banking entities, each accompanied by its own set of advantages and drawbacks.

A pivotal decision rests upon understanding the long-term trajectory of your enterprise which is easier said than done as a financial technology company.

Selecting the correct point of integration holds profound ramifications, influencing your scalability potential and transactional margins within the partner ecosystem.

When considering your integration, a clear sense of direction is crucial. Picking the appropriate entry point is akin to selecting a life path. It can be a determinant of success or an impediment to growth. An ill-fated choice may undermine your ability to scale and optimize transactional efficiency within your chosen partner constellation.

The accompanying table serves as a compass, delineating various integration options into the payments and financial services supply chain:

Entity Type	Integration Type	Supply Chain Location	Terms Used
National Networks	ISO 20022, ISO 8583	Wholesale - Network Source	FedNow, RTP, Visa Direct, MasterSend
Regional Networks	ISO 20022, ISO 8583, APIs	Wholesale - Network Source	NYCE, STAR, PULSE, SHAZAM, Mastro

Traditional Fintechs & Core Providers	ISO 8583, ISO 20022, ACH, Batch File Based (APIs are limited)	Wholesale	Core Platform, Merchant Acquiring Processor, Card Issuing Processor, RTP
Bank	API, Reports, Batch File Based (Varies based on Bank maturity)	Direct with Bank	Open Banking, Sponsorship, BaaS, Embedded Payments
BaaS Provider	API, Reports, Batch File Based	Payment Service Provider (PSP) Fintech with Sponsor Bank or Fintech licensing software only	Open Banking, BaaS, Payment Provider
Embedded Finance	API, UI Portal, Reports	Payment Service Provider (PSP) Fintech with Sponsor Bank	Embedded Payments, PayFac, Payment Provider

Card or Lending Product	API, UI Portal, Reports	Payment Service Provider (PSP) Fintech with a Sponsor Bank Prepaid, Credit, Debit & Loans	Card Issuance, Prepaid, Credit & Debit Issuance
Merchant Services or ISO (Multiple terms are used in the market such as Gateway, Hub PayFac, etc.)	API, Reports, Various Interfaces, UI Portal	Payments Company Connected to the Payment Service Provider (PSP) Typically not connected directly with the Bank. Indirect with various providers.	BaaS, Embedded Payments, Niche Markets, PayFac, Payment Provider

It is important to note that the last (3) rows in the table are variations of Merchant Service Organizations, all with various degrees of separation between the merchant or end customer and the bank. Organizations that support B2B transactions or B2B2C models utilize different integrations at different points of entry, much like the Independent Sales Organization (ISOs) in the credit card processing ecosystem, while from a regulation standpoint they are strictly sales organizations with no technologies of their own but in today's landscape they tend to add technical solutions to help differentiate their product and services. These areas are continuously evolving with the market changes and regulatory hurdles.

When connecting to banking providers, it is important to realize that banks have varying degrees of maturity as it relates to technology. Many have APIs that you can connect to them with, while others are largely dependent on their core provider and their core provider's stance on allowing access to their system, which is largely why Banking As A Service (BaaS) providers exist.

Comprehending the Layers

In our pursuit of a deeper understanding of the financial landscape, it is crucial to focus on the top two rows of the table above, which encompass the Wholesale facet of banking. This sector in between the networks and the bank plays host to some of the world's largest fintech corporations, offering a wide array of software products and services. These offerings span from core platforms to debit card and payment solutions, all intricately connected within the wholesale networks.

In addition, organizations such as Bankers Banks, Liquidity providers, InterBank ACH Providers, Correspondent Banking Networks all live in this space of the supply chain.

It's worth noting that in recent years, certain entities have defied conventional categorization and have skillfully traversed supply chain boundaries to engage directly with consumers and businesses. Amidst these evolving dynamics, our primary objective remains to provide clarity and guidance on the various pathways for integration by keeping in mind, as the networks acquire new fintechs, this will continue to evolve who offers what services.

The current market has put the bank in a different position from in the past, whereby the banks are now plugging in where necessary based on their market and risk appetite and have become less of the gating factor even though by law they are always required to be in the flow of any monetary movement

In the realm of business, this landscape resembles a bustling highway for financial transactions, akin to a steady stream of vehicles, each participant anticipating their toll.

As you navigate through the layers of integration and what point of entry you select, the decision really comes down to what are you trying to accomplish and what niche expertise and value-added services will you provide to your customers, understand, the father you are down the supply chain the higher the cost since there are more tolls to pay the providers in between you and the bank, potentially inflating the expenses associated with your program.

Crafting a Clear Product Blueprint

When selecting the most suitable partner for your payment program, a pivotal factor hinges on the alignment of your intentions with your chosen provider. Whether you're a business seeking to seamlessly integrate payments or a startup intrinsically linked to money movement, the roadmap to informed decision-making revolves around your use case and the related laws and regulations.

In short, define your product scope, diagram the workflows and funds movement and clearly define the roles and responsibility that you envision your partners to have and how you will measure and manage their performance. Your providers in your program are basically Third Party Providers which require a set of standard operational procedures for vendor management so that you can manage they are meeting all legal and regulatory compliance requirements.

While the notion may appear elementary, grasping the essence of the customer's journey stands as a paramount endeavor so you can both ensure a low friction experience but also encompass the array of validation checkpoints and systemic safeguards for fraud and security, some of which may be provided by third party service providers. This robust orchestration ensures regulatory compliance and fosters a seamless customer experience.

At the core of your product definition lies the design of a user experience (UX) that encapsulates both functionality and privacy considerations while protecting the customer and transaction data from bad actors.

Privacy, an indispensable facet of modern digital interactions, assumes a prominent role. As part of your product definition, considerations must be extended to accommodate privacy constraints and consumer protection laws outlined above, ensuring that user data remains secure and aligned with evolving data protection standards.

Furthermore, your product's framework requires the integration of real-time interactions with external third-party providers within the customer onboarding process. It also needs to be part of normal account operations, since user changes and updates could expose fraudulent behaviors. Considerations for transaction speed, exception management and unhappy paths are a critical component and indispensable for meeting Know Your Customer (KYC) and other compliance requisites. Every step along this journey must resonate with legal, operational and compliance obligations.

In essence, the realm of product definition is a multi-dimensional canvas where aspirations meet practicality. Your blueprint not only outlines your program's purpose and objectives but also delineates the unique workflows of customer engagement, compliance adherence, and seamless user experiences.

Technical Orchestration

When building a payments platform, your technical architecture and platform construction typically will focus on the seamless integration of data streams and the assembly of components to bring the complete lifecycle to fruition within a delightful customer experience. Nevertheless, for those unacquainted with the complexities of building a transactional processing engine, there exist critical aspects worth exploration. The following are a couple notable subjects that can help to avoid finding oneself metaphorically backed into a corner.

Contemplating an Apocalyptic Scenario

Beyond mere jest, consider this scenario: if a catastrophic event were to unfold today, would your system endure the turmoil, standing resolute for your valued customers and banking sponsor?

In the dynamic landscape of any business, marked by shifts, pivots, and ever-evolving market forces, the infrastructure you establish must possess the resilience to adapt, evolve, and expand in harmony with your company's clientele, its partners, providers and banking relationships.

Consider a deep examination of the concept of the "System of Record (SOR)". Instill the principle that even in the event of disconnection from third parties or external entities, your system should continue to function autonomously. Such a posture fortifies your operational continuity, ensuring that your system's uptime remains unwavering, irrespective of external dependencies.

This approach also will help provide a clear path for Disaster Recovery (DR) and Business Continuity Planning (BCP) documentation and policy controls. It will also help you best understand your liability or risk associated with loss of connectivity and possible validation efforts.

Establishing the System of Record (SOR)

The establishment of the System of Record (SOR) entails designating a data repository as the authoritative source for specific data elements. This repository holds the creation, modification, updating, and deletion of data objects or their attributes. This custodianship forms the bedrock for regulatory reporting, an indispensable facet underscored in previous discussions on compliance.

Aside from Personally Identifiable Information (PII) and other protected data, this repository should include the relevant meta and log file data that allows you to recreate each and every transaction related to the end customer. The question you should ask yourself is, if something happened, can you prove that it happened the way you have it documented in your policies and procedure documentation? If you can't prove it, it didn't happen. If you design your data repository with that in mind, you will be prepared or recover from both catastrophic outages and regulatory audits.

Architecture Design

As you embark on your architectural journey to build your platform, deliberate upon these fundamental design principles that should shape your platform:

1. Multi-threaded transaction processing

2. Modular architecture for scalability

3. Simplified technical orchestration for scalability

4. Real-time redundancy for enhanced reliability

5. Warm-hot vs. hot-hot environments

6. Data repository backups environments

7. Recognizing security's distinct realm from compliance

8. Financial and compliance reporting

9. Evidence creation for compliance adherence

10. Business Key Performance Indicators (KPIs)

11. Prompt exception notifications

12. Data segmentation strategies

13. Vigilant fraud identification, tracking, and management

14. Aligning customer complaints with compliance and fraud monitoring

15. Separating customer onboarding from operational management

16. Ensuring user-friendly UX aligns seamlessly with embedded compliance

17. Championing robust operational repeatability

18. Disaster Recovery (DR)

19. Business Continuity Plan (BCP)

20. Continuous Integration (CI) Continuous Delivery/ Continuous Deployment (CD) Processes

For a start-up, some of the topics above are daunting and outright cost prohibitive on day one, but the idea is to give you an understanding of what you should be doing. Once you have started transactional processing, some of the initiatives above can be developed over time; but regardless should be part of your product and technical roadmaps.

Embracing Third-Party Dynamics

The financial services arena pulsates with the presence of third-party entities performing pivotal functions such as Identity Validation, Account Validation, Technical Orchestration, and even Auditing. These entities play a critical role in fulfilling a company's regulatory responsibilities spanning Customer Onboarding, Fraud Monitoring, Security, and Compliance.

Operating within these spheres entails managing vendors and having a defined Third Party Risk Management program whereby you can manage your third party providers through effective risk management practices while validating their contractual and regulatory performance.

Beware of Embedded or Nested Processing

Embedded processing emerges as a potential pitfall. Inadvertently venturing into this territory poses regulatory hazards, as it can obscure the transparency banks require to manage Anti-Money Laundering (AML) risks.

Nested processing is when the business customer or otherwise a merchant is also a payment processor, resulting in a "nested" payment processor or "aggregator" relationship.

This restricts the visibility of the data and who the end customer actually is which creates a blind transaction whereby the company could be ineligible or a restricted organization that isn't regulatory compliant.

This is why crafting Third Party Risk Management strategies becomes pivotal, enabling clear oversight and transparency. As vendors for your organization, each third-party entity warrants diligent due diligence before integrating their services into your production environment. These measures safeguard your business and align with the overarching framework of regulatory compliance.

Building a Compliant Operational Framework

In the dynamic realm of startups, where the spotlight often shines on sales as the key to success, operational considerations sometimes find themselves in the shadows. However, astute entrepreneurs understand that a strong operational foundation is the cornerstone upon which sustainable growth is built and put it simply, allows you to scale in a formalized compliant environment.

Readily available and reliable customer service coupled with a set of standard operational procedures is another key to building and sustaining client and partner relationships. One misstep on either of these could cause irreparable relationship situations that could take a very long time to rebuild as word travels fast across the industry and over the internet.

Standard Operating Procedures (SOPs)

Steering your operational ship and preparing to scale mandates the implementation of Standard Operating Procedures (SOPs)—a framework ensuring uniformity, efficiency, and compliance of repeatable processes. Each SOP serves as a guiding compass, leading your team through multifaceted operational tasks. It's akin to a military playbook deeply ingrained in your operational DNA, streamlining your approach and minimizing miscommunication.

Operational Processes

Consider your SOPs as your centralized location for many of your controls and how you will create and provide evidence of your compliance with regulations, fraud protection and commitment to security and customer data protection, and of course a clear guide for your operational process for new employees and third party auditors alike.

When developing operational procedures break controls into the following sections;

- Informal/Undocumented

- Formal/Written & Documented

- Documented Data Flows

- Documented Transaction Flows

- Automated System Processes

- Manual Processes

- Processes using a Hybrid approach

As a growing company it is hard to always have everything clearly documented, While both banks and regulators understand this, you should illustrate that you are aware of the processes and how they help control the risk within your processes even if they are not documented.

Review Processes

In the world of banking there is a phrase used to describe the relationship a bank has with their customers and it's *"Trust & Verify"*

Consider the second half of that phrase, the purpose of this section, which is to allow your sponsor bank, regulators, third party auditors and your team to Verify your controls and your processes are effective.

The following operational processes are best practices to have in place to test and monitor your procedures. It is important that you consistently validate that your control processes are up-to-date and compliant with your policies as you have defined them.

1. **Internal Audit**

Perform internal audits of your procedures by conducting reviews such as;

- Document Reviews
- Transaction Testing
- Employee Interviews
- Mystery Shoppers

2. **External Audit**

Non bias third-party audits are the highest form of validation of your procedures and is what most banks and regulators prefer

3. **External Certification Assessment**

There are only a few available Certification Assessments that actually provide a certification through third party providers for operational processes specific to the banking industry that would relate to a Fintech or Service Provider. The most commonly known are;

SOC 2 Type II (Operational Process Validation)

The SOC 2 Type II is an assessment of the operating effectiveness of your internal controls over a period of time, typically 6 -12 months. SOC 2 Type II audits require a greater investment of both time and resources and require at least 6 months of history to validate. This certification can take from 9 to 15 months to acquire.

Standardized Information Gathering (SIG) ie. SIG Lite (Security Process Validation)

This Assessment is designed to provide a broad, but high-level understanding of internal information security controls. The SIG Lite is for organizations that need a basic level of due diligence assessment.

The SIG Lite assessment questions allow organizations to simplify and standardize their third-party risk management and compliance initiatives by aligning multiple controls and regulatory requirements across various security frameworks.

Payment Card Industry i.e. PCI (Payment Card PII Process Validation)

The Payment Card Industry Data Security Standard (DSS) is an information security standard used to handle credit cards from major card brands. The standard is administered by the Payment Card Industry Security Standards Council, and its use is mandated by the card brands. So if you accept, collect, transport or store payment card information, this will apply to your organization.

4. Quality Assurance / Quality Control

Quality Control is more than just QA software development testing; it is a commitment to the validation of the "intended outcome". It includes reviewing change management processes, having effective roles and responsibilities with the appropriate access rights and having processes that ensure both systemic and human resources are delivering as expected.

Quality Assurance is the gating process ensuring you are producing the high level of quality of service, product build and of course customer satisfaction. So doesn't it make sense you want to review these procedures and monitor that processes are as effective as possible.

5. Key Performance Indicators (KPI)

Every business needs KPIs to track their performance, KPIs in Fintech tend to always be focused on revenue and customer acquisition and not prioritize the metrics that allow you to illustrate your effectiveness with how well you run your operations and most importantly your controls that manage data security and regulatory compliance.

By prioritizing operational metrics, it both provides you with direct insights to possible problems with delivering your service to customers and potential fraudulent activities. It also provides you with an opportunity to illustrate your operational effectiveness to your external partners.

Establish metrics that illustrate your performance in the following areas;

- System Performance
- Compliance Performance
- Exception Reporting

- Fraud Management

- Customer Service

In today's market, not enough leaders place importance on compliance, fraud management and operational processes. Without them you may find yourself performing an emergency pivot or even out of business when your banking sponsor severs your relationship due to their concerns with your ability to effectively protect your customer's information or even just the bank's brand image.

Never forget, there are only two ways to legally move money in the United States. (1) Obtaining Money Transmitter Licenses (MTLs), at a cost of almost 2 million dollars and two years of time. (2) Obtaining a Banking Sponsor. So keep your Bank Sponsor happy and out of trouble with the regulators!

Training

Training is a requirement in almost every single control framework. Regardless if the subject is Data Protection, Security, Operational Procedures or even Customer Service, training is a critical function to fortify a strong process and frankly a required compliance activity. Build your training programs for the entire organization as well as for specific teams of groups to narrow down the message that are unique to various roles and responsibilities.

- General employee training

- Custom tailored team training

- Board of Directors Awareness Training

- Third Party & specific stakeholder training

Contractual Obligations

Within every contractual agreement, there are a series of legal obligations you have as an entity. In the banking and fintech industry it is typical for many of the clauses related to Data Security, Consumer Protection and Regulatory Rules & Laws to be cascaded down to your partners, distribution stakeholders & third-party payment providers.

This practice serves two purposes, reduction of liability and third-party enforcement of regulatory compliance responsibilities and best practices.

Regulatory compliance flow through illustration

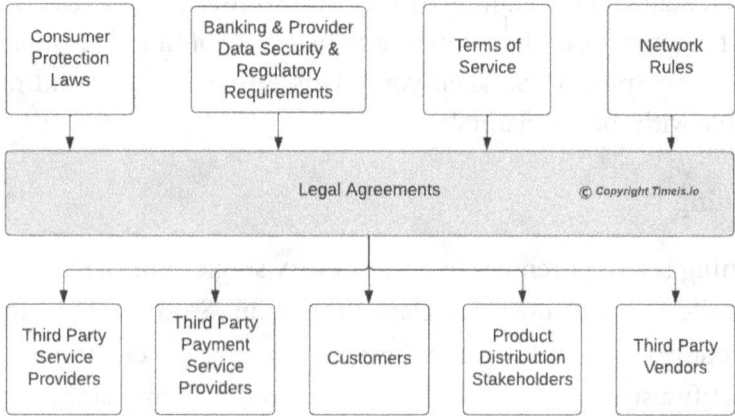

Consumer Protection Laws	Banking & Provider Data Security & Regulatory Requirements	Terms of Service	Network Rules

Legal Agreements
© Copyright Timeis.io

Third Party Service Providers	Third Party Payment Service Providers	Customers	Product Distribution Stakeholders	Third Party Vendors

As the diagram above illustrates, as a financial technology provider or even a corporation looking to enable various forms of real time payments, realize you are a cog in the wheel in the larger ecosystem of financial services.

As you take your place in the logistical supply chain of payments, you will become not only a service provider for your customer but also a vendor manager for all the stakeholders and partners in the lifecycle of your process in the delivery of your product or service.

So when you are contemplating building relationships with any one company, know that in some cases you could create difficult legal obligations on both sides of the customer - provider dynamic.

In other words, customers you typically are focused on pleasing and giving them what they want or need, while vendors you tend to have a different relationship with whereby you need to demand a level of service and support, establish Service Level Agreements (SLAs) and hold to what they say they will do.

The important point to make here is to be aware that this type of relationship could put yourself in a "conflict of interest" especially if the relationship is tested when things go wrong and often become more complex and challenging to manage.

Employee Obligations

In a regulated environment, all employees have a role to play in meeting corporate compliance and effectively defending your organization against fraud.

Aside from the standard participation in company wide training on security and compliance practices, they become your front lines of defense against malicious actors and protecting your customers' funds, information, and privacy.

Roles and responsibilities place a key role on helping your employees best understand their role in the process of defending your organization and protecting your customers. It also acts as a control point by creating clear segmented roles to reduce the opportunity of internal collusion or any one employee getting leveraged or manipulated by outside forces.

Creating separate roles that allow for you to establish rules such as the Maker - Checker control will help reduce the risk of a single employee being compromised and/or performing actions that could result in increased risk, liability for fraudulent activity for the company.

Risk Management & Compliance

Risk Management and compliance is an integral element that should be intricately woven into your business model. The processes should not only scrutinize your operational & legal processes, but it also validates your commitment to regulatory compliance, most importantly for the benefit of your company, its customers, investors and your direct or indirect bank sponsor.

Compliance goes beyond a mere collection of documents; it represents a series of controls that embody regulatory rules, consumer protection laws, network rules, best practices and standards. These controls serve as a protective shield, safeguarding customers, your company, the sponsor bank, and the broader financial ecosystem—a vital defense against systemic risks within Fintech and Banking relationships.

Severe penalties as well as irreparable reputational harm can result from any compliance failures. So please take it seriously so you can best protect the house you are building and its customers that live there.

Establishing a Risk Management approach to managing your business, will not only help you understand how regulators and your banking sponsor views your business model, but it will also help you prioritize what and how to manage the inherent risk in your business.

You can never mitigate risk in your business model, but you can do things to manage the risk. Below are a number of operational efforts you can undertake to help manage the inherent risk in your business.

Embedded Compliance Controls

Just as in product development, the challenge of compliance demands a deliberate approach to focus both on the customer's journey and user experience all while establishing defenses against bad actors and malicious characters attempting to do harm.

It will be helpful to outline a comprehensive roadmap of requirements that align your reporting requirements with regulatory outputs. In addition to helping illustrate your conformity with regulation, it will also help you think in terms of what data is needed to tell the story of your customer lifecycle through onboarding and payment transaction workflows. Remember if you can't produce data that illustrates you have a control in place, say to validate a user's identification, then in the view of regulation it didn't happen. Proof is in the delivery of evidence that you did what your documented policies say you do.

While the process can be time-consuming, the experienced hand of an advisory practitioner can expedite the path, guiding you to the evidence that best attests to your compliance efforts.

Central Compliance Pillars

Fintech startups, while often preferring flexibility, must cultivate a repeatable operational structure that meets regulatory requirements. Defining and documenting the following pivotal areas will help you build a structured approach that is expected.

Area of Focus	What's Required	Intent
Customer Identification	Written policy & operational controls supporting KYC - KYB & beneficiary owner validation	Validation of "Who is" the user, customer and / or device. *Question you are answering;* **"Are you allowed to go through the onboarding process?"**
Customer Onboarding	Written policy & operational controls supporting KYC - KYB & sanctions	Validation of "Who is" the account holder, user or customer and saying "No" to users that are on the bad actor lists. *Question you are answering;* **"Are you allowed to open an account, have access to those** **accounts or funds and meet the criteria of our product or service?"**

Anti Money Laundering (AML)	Written policy & operational controls managing & monitoring, sanctions & who, how and where funds are moved.	Monitoring and controlling processes to keep bad actors out of your system or using your system to send or transfer funds, especially funds derived from illegal practices. Many fraud controls are defined as part of this process since account take over and other phishing scams are ramped in today's environment where your customer ends up as a proxy for the bad actors. Question you are answering: *"Is this transfer of funds legit and for the intended purpose by our validated customer?"*
Incident Reporting	Written policy & operational procedures that meet your business model's regulatory requirements and best practices.	In the case of a material event such as a data breach, hack or even system outage what is the process for addressing the issue and communicating with various stakeholders. Question you are answering: *"Where are the transaction exceptions and negatively impacted events in the process of identification, investigation, resolution and communication at any one point in time and how do you notify and disclose."*

Investigations	Written policy & operational procedures for addressing fraud and potential fraud.	A formal process for addressing and investigating customer and transaction behavior to identify and manage the risk of various forms of fraud Question you are answering; *"Here is how we effectively manage and resolve fraudulent behaviors* *and actors utilizing our products and* *services."*
Customer Notification	Written operational procedures supporting customer disclosures.	Keeping Customers aware of possible impacts to their accounts, funds or Personal Identifiable Information (PII) Question you are answering; *"How we notify and communicate with customers of material events or* *required regulated notifications."*
Information Security (InfoSec)	Written policy & operational procedures outlining controls for data protection, retention and security protocols	Data protection for your system, product, service and customer data from external hackers, internal compromised employees and other bad actors. Question you are answering; *"How effective are our data protection, defensive controls and* *procedures"*

SOC 2 Type II	Written operational processes across all parts of the business defining internal controls and systems related to security, availability, processing integrity, confidentiality, and privacy of data.	Internal repeatable operational controls and systems that will allow for the business to operate and process transactions with integrity and confidence while effectively managing data privacy and security protocols Question you are answering; *"How well we consistently process transactions reliably and control access to various forms of protected information"*
Customer Complaints	Specific review processes for identified potential fraud.	Identification of potential fraud and anomalies in behaviors that may indicate fraud. Question you are answering; *"How we review Customer complaints for potential fraudulent activities."*

Third Party Vendor Management	Written policy & operational procedures outlining controls for Third Party Risk Management (TPRM) & Vendor Management	Manage the systemic risk associated with third parties that are involved in the delivery of your product or service. This includes the cascading series of regulatory compliance for data protection, data privacy and customer protection laws <ins>Question you are answering;</ins> *"How well we manage our partners and stakeholders in our business* *model to adhere to regulations and compliance requirements"*
Proof of Compliance	Provide proof of compliance controls and fraud defensives	Validate you are actually performing as you have stated in the companies' policies and procedure documentation. <ins>Question you are answering;</ins> *"How well you are at self monitoring your processes and have a handle on if your controls are effective"*
Data Structures	Be in position to track, monitor and report on your entire customer lifecycle while illustrating the controls you have in place.	Reporting and Compliance KPIs <ins>Question you are answering;</ins> *"How well you are at self monitoring your processes and have a handle on if your controls are effective"*

Settlement & Reconciliation	Tracking and Validation of all funds moving in and out of the company's payment processes.	Balance and reconcile the payment processes and account balances to the penny. <u>Question you are answering;</u> *"How well you manage the flow of funds in and out of your business model and partner bank accounts"*

This list of operational processes and procedures serves as evidence of your culture of compliance, weaving an intricate tapestry of controls and safeguards for your organization. Along the way, producing the required evidence of successful controls throughout the customer's journey addressing these processes and procedures will only help you stay regulatory compliant and most importantly protect your customer's information and money from bad actors.

Conclusion

The implementation of instant payment functionality can be daunting and seem overburdened with complex compliance and regulatory requirements. While there is truth to the amount of compliance vectors that need to be considered based on your business model, if approached in an orderly fashion it can help you build a solid foundation for your business and ensure you evade a number of materially catastrophic events that could put you out of business in the future.

The following are a couple simplified key takeaways from this chapter;

- Know your customer, your market and their characteristics

- Know your business model's regulatory requirements

- Pick partners and providers wisely

- Validate, monitor and test customers, changes and processes continuously

- Document what and how you do things

- Be in position to prove your documented controls are being performed.

- Build a platform that is modular and that can pivot and scale at the same time

- Build your product or service with strong data collection and compliance controls built in

- Done right, your compliance and risk management processes will be a precursor to your ability to scale your business and retain your banking relationships.

Chapter 5 - Introduction to instant payment use cases and their benefits

By Kevin Olsen

In the beginning

"If I had asked people what they wanted, they would have said faster horses." Henry Ford

Mr. Ford was speaking about the innovation of the automobile, and the automobile was a drastic change from the horse. In the case of instant payments in the U.S. via FedNow® and RTP®, we can deliver what people want, the faster option, and introduce new products and services.

The question then is, "Are you ready to start using instant payments?" You may be wondering where and when I would use instant payments. Knowing the advantages to consumers, businesses, and financial institutions helps, but where do you start, where do you look, and how do you create use cases that will be useful?

The shift to digital channels has led to the popularity of instant payments. It has accelerated the shift to online and mobile banking, as consumers increasingly prefer to manage their finances through digital channels[3]. This trend has led to a decline in the number of physical bank branches, reducing the number of potential targets for robbers. According to the FDIC, the total number of bank branches in the United States fell by more than 3% between 2017 and 2020[13].

One of the primary reasons for the decline in bank robberies is the global transition from cash-based transactions to digital payment methods. Instant payment systems, such as Zelle in the United States, Faster Payments in the United Kingdom, and the Unified Payments Interface (UPI) in India, have revolutionized how people transact, making it quicker and more convenient than ever before[31]. A 2019 Federal Reserve report showed a decline in the use of cash for payments, with digital payment methods such as debit cards, credit cards, and mobile wallets gaining popularity[13]. The COVID-19 pandemic further accelerated this trend, as people opted for contactless payment methods for safety reasons[11].

When discovering use cases for instant payments, one of the first areas to consider is the approach you will take, and a highly effective starting point is the philosophy of "Alternative and Addition." This methodology involves scrutinizing your existing payment channels, products, services, and other offerings to identify pain points and create value. Specifically, you can explore areas where instant payment solutions can serve as an alternative in some cases to your current offerings, and in other instances, act as an addition to your existing options[17].

Help me with my Accounts Receivable Automation

Whether they're utilities, insurance, mortgage, lenders, or healthcare providers, all billers have at least one thing in common: accounts receivable (A/R) processes that have moments where things go wrong, and a life preserver is needed. Instant payments is that life preserver and more.

Regardless of the industry or sector, organizations that deal with accounts receivable (A/R) processes often face challenges and require support for Straight-Through Processing (STP). Instant payments can provide the necessary support and more to address these issues. Payment-related complications can drain resources, often requiring significant personnel investments to resolve. If this kind of problem often arises in an organization, it might be time to look at how things are done now and consider using instant payments for Straight Through Processing (STP)[8].

To better understand how instant payments can be beneficial, consider some examples of where it can provide value.

There are issues with payments where organizations are forced to dedicate many resources, usually people, to resolve those issues. If you have situations with external businesses or internally where these issues occur more frequently, that's the place to revisit the process changes[23]. Instant payments have Straight Through Processing (STP) opportunities. Now, let's move on to examples of areas where instant payments can help.

"Please don't cut me off" Bill Payment

Living in Florida, it is imperative for me to keep my air conditioning on, especially during the hot and humid month of August.

Additionally, maintaining a solid credit record is crucial. Let's consider a hypothetical scenario where I make a payment for my electric bill via check. Still, due to the slower mail system, the payment gets delayed, resulting in a late fee. If instant payments were available, I would have the flexibility to make the payment up to the actual time payment is due, rather than just the due date, affording me extra time to arrange funds or transfer them to the appropriate account. Using instant payments can significantly reduce late payment fees and disgruntled customer calls[30].

The benefits of instant payments extend beyond just utilities, as it can easily be used for other payments such as rent, mortgage, and even car loans[2]. With instant payments, the risk of eviction, car towing, or uncertainty surrounding payment dates can be eliminated. Unlike other payment options, Instant payments provide immediate payment confirmation. While this confirmation doesn't always equal successful posting, in most situations this means customers would call the biller's accounts receivable department less often to find out if their payments have been received and processed[15]. This results in a win-win scenario for all parties involved.

Also, instant payments eliminate the chance of payments being sent back, removes NSF seen in ACH processing and insufficient card balances experienced in card payments which is a major benefit for businesses[16]. The immediate payment confirmation also makes customers feel safer and more confident in their payment methods, which leads to happier and more loyal customers in the long run[15].

Instant payments offer numerous benefits to consumers when paying bills, such as reducing late payment fees, providing payment flexibility, eliminating returned payments, and offering immediate payment confirmation, all of which increase customer satisfaction.

Are you going to return that? - Return Processing.

Occasionally, customer payments may be reversed, returned, or charged back. This occurs due to various reasons, such as inadequate funds in their checking accounts, exceeding their credit card limits, expired cards, or even account closures. Recurring returns can pose a significant challenge to businesses, especially those with a high volume of transactions. However, with instant payments, the likelihood of such occurrences can be significantly reduced.

Instant payment's irrevocable "push" payments ensure that account holders can only make payments when they have sufficient funds available in their accounts, although institutions have the choice to allow payments to be sent beyond what is available and dependent upon their software abilities[14]. As such, the risk of returned fees due to insufficient funds or over-the-limit card balances is greatly reduced. In contrast to other payment methods, where payments can be easily reversed or charged back, instant payments ensure that funds are transferred securely and reliably.

With instant payments, businesses can benefit from the added assurance that payments will not be returned due to insufficient funds or over-the-limit card balances[8]. Additionally, organizations can save significant resources by avoiding reconciling returned payments or dealing with chargeback requests. However, it is essential to have a mechanism to facilitate disputes via request for payment returns that consumers rely on when making purchases. Instant payments do offer a mechanism for the request for return for funds, offering an alternative to handling costly and risky chargebacks, a burden typically managed by banks. By leveraging these instant payment capabilities, companies can streamline their payment processing operations and provide their customers with a secure and reliable payment option. This raises the question and demonstrates how the benefits of using instant payments, such as reduced chargeback costs, can outweigh banks' interchange revenues, potentially encouraging them to support use cases that primarily generate interchange fees.

"Save the Paper" - Electronic Billing

Invoicing is an integral part of the billing process, and it can also be a challenging one. Accuracy and timely payments are essential for maintaining good customer relationships, but many variables are involved in the invoicing process that can lead to errors or delays. Instant payments can help address these issues and streamline invoicing.

One way that instant payments can be used for invoicing is by including a Request for Payment (RFP) message with the invoice[18].

This message can document all of the payment information, such as the account number and itemized details about the bill, and the customer can pay right away using instant payments. This approach ensures that payment is made promptly, reduces the chance of errors or missing information, further reduces the number of disputes, and gives customers greater control over the payment process[20].

Another benefit of using the RFP process within instant payment for invoicing is that it can help businesses take advantage of early payment discounts[20]. Organizations may choose to provide a link to the bill or invoice as well as incentives for early payment. Companies can incentivize early payments and improve their cash flow by offering customers a clear payment window and instant payment options. This method can be helpful for businesses with small profit margins or little cash on hand because it lets them better manage their payments.

Moreover, invoicing with instant payments also can help reduce the administrative burden of managing invoices and payments[9]. By automating the invoicing process, businesses can eliminate the need for manual data entry, reduce the risk of errors, and free up staff time for more value-added tasks. This method can help businesses improve their operational efficiency, cut costs, and focus more on their core business tasks. With instant payments, businesses not only gain the assurance that payments will not be returned due to insufficient funds or over-the-limit card balances, but they also stand to benefit in the automation of their Accounts Receivable and Straight-Through Processing (STP) processes. Instant payments can deliver electronic remittance data alongside the payment, enabling automated reconciliation and thereby streamlining the entire payment lifecycle.

Instant payments provide businesses with a powerful tool to support their invoicing and payment processes. By including Request for Payment (RFP) messages with invoices, organizations can offer their customers greater control over the payment process, reduce errors, and improve their cash flow[18]. With instant payments, businesses can streamline invoicing processes, reduce administrative burdens, and focus on their core operations.

Are You Going to Fix that Invoice?

Have you ever had a situation where a customer needed to provide accurate or complete remittance information with the payment?

It leaves the A/R department guessing which account and which invoices, credit notes, and other adjustments to post to when it happens. If the wrong account or invoice is credited, the A/R department may end up re-invoicing and possibly charging late customer fees or even denying an early payment discount[8]. This is "no bueno" in customer service.

Intensifying on the other side, the customer's accounts payable (A/P) department may pay the duplicate invoice, resulting in overpayment issues that must also be resolved. Side note: The Request for Payment (RFP) functionality serves as an effective mechanism to mitigate the risk of duplicate payments or fraudulent invoicing. Within instant payments networks using RFP's, each payment request is associated with a unique reference number, making it easier for identifying potential duplicate invoices or payment requests. This unique reference number helps to ensure that each RFP is only paid once, thereby preventing accidental or unintentional overpayment. If a payment request with a duplicate reference number is detected, the system can automatically flag it for review or block it altogether. This provides an additional layer of security and auditability, enhancing both the payer's and the payee's confidence in the payment system. Overall, the RFP functionality with its unique reference numbers significantly contributes to the efficiency and integrity of instant payment systems using the RFP mechanism.

In any case, whether you are AR or AP, this is a pain that can be relieved with instant payments. Billers can reduce the incidence and costs associated with incomplete or missing remittance information by sending customers an RFP message via instant payments. The notice includes the account number plus pertinent invoice details, enabling the automatic inclusion of the remittance information when the customer makes the payment.

Want to take it a step further? Think of it as sending an e-invoice. An invoice that accounts payable could upload into their A/P system and automatically schedule payments to take advantage of any early payment discounts or pay on the due date in real time to maximize working capital and liquidity[1].

It will be exciting to see what other solutions we can develop to reduce returns, improve just-in-time payment options, and make life easier for accounts payable and receivable.

Request for Payment (RFP) with instant payments - Breakdown by participant

Business-to-Business (B2B) payments

One of the most significant benefits of RFP technology is its ability to streamline B2B payment processes. This technology has the potential to revolutionize the way businesses receive payments, allowing them to improve their cash flow management and overall financial position[2]. With RFPs, companies can ask their customers or clients for safe and timely payments, ensuring that the money is received quickly and easily. The RFP functionality can also be designed to specify the type of remittance advice that should accompany the payment. This facilitates automated Accounts Receivable and Straight-Through Processing (STP), as the electronic remittance data can be easily matched with the corresponding invoice. By embedding this information directly within the payment process, organizations can reduce or eliminate manual reconciliation efforts, reduce errors, and expedite the entire Accounts Receivable process.

RFPs can be especially valuable for businesses relying on a steady stream of incoming payments to maintain operations. By utilizing this technology, companies can reduce the time it takes to receive payments and avoid delays caused by traditional payment methods[16]. This helps ensure that funds are available when needed and helps to prevent cash flow issues that can negatively impact a business's financial stability.

Another significant benefit of RFPs is the cost savings they can provide. Traditional payment methods, such as paper checks and wire transfers, can be expensive and time-consuming for the payer and payee[17]. By contrast, RFPs are relatively low-cost and can be processed quickly and easily, reducing the time and expense associated with traditional payment methods.

Overall, RFPs represent a significant improvement over traditional payment methods for B2B transactions. They offer greater convenience, speed, and security while reducing costs and improving cash flow management. As more businesses adopt this technology, RFPs will likely become an increasingly common and essential component of B2B payment processes[18].

Professional Service Organizations

RFP invoices can significantly benefit service providers and the many organizations they serve beyond financial institutions, including freelancers, consultants, and independent professionals. By sending RFP invoices to clients, service providers can receive payments quickly and efficiently, eliminating the delays associated with traditional invoicing and payment methods[18].

RFPs enable service providers to request payments from clients in real time, improving their cash flow management and helping to reduce the time it takes to receive payments. Additionally, RFPs can help service providers reduce the costs associated with traditional billing and payment methods, such as paper checks and wire transfers[2].

This level of convenience is essential in today's fast-paced business environment, where service providers often need to get paid quickly to ensure they can maintain their operations and maintain client relationships. By incorporating RFPs into their billing and payment processes, service providers can simplify the payment process for their clients and reduce the administrative burden on their organizations.

Furthermore, RFPs can help improve service providers' overall customer experience. Service providers can improve customer satisfaction and loyalty by offering request for payments and real-time payment options, which can translate into more business opportunities and growth[18].

RFPs can benefit service providers significantly, allowing them to receive payments quickly, efficiently, and securely. By leveraging this technology, service providers can improve their cash flow management, reduce costs, and enhance the overall customer experience.

Retailers

The widespread use of digital platforms has revolutionized the retail industry, making it easier and more convenient for customers to shop and make payments online[4]. Because of this, stores and online businesses need to use advanced payment systems to speed up the checkout process and make the whole customer experience better.

One such payment system that has recently gained prominence is the Request For Payment (RFP) system. RFPs can be used by retailers and eCommerce businesses to ask customers for payments, especially when customers make purchases online using their smartphones or computers[18]. The RFP system allows these businesses to send payment requests to their customers in real time, enabling them to complete their transactions quickly and securely.

By using RFPs, retailers and eCommerce businesses can make the payment process easier and reduce the chance of mistakes that can happen when payment requests are made by hand while also providing additional information related to the purchase. This streamlined payment approach also can help companies to minimize operational costs, as they no longer need to invest in time-consuming manual payment processing methods[2].

Also, using RFPs can significantly improve the overall customer experience by making the payment process smooth and easy. Customers can respond quickly to payment requests sent through RFPs and finish their transactions quickly, which makes customers happier.

RFPs are a new and efficient way for retailers and businesses that do business online to ask customers for payments. By utilizing RFPs, companies can enhance the customer experience, improve operational efficiency, and increase customer satisfaction, leading to long-term business success.

Nonprofits and charitable organizations

Nonprofit organizations and charitable institutions depend primarily on the generosity of their supporters to fund their operations and initiatives. The Request For Payment (RFP) system has emerged as a powerful tool for nonprofit organizations to request donations from their supporters[2].

With RFPs, nonprofit organizations can send donation requests to their supporters in real time, enabling them to receive donations quickly and securely. This streamlined approach to fundraising can help nonprofit organizations improve the speed and efficiency of their donation collection efforts, allowing them to focus on their mission and objectives.

By leveraging RFPs, nonprofit organizations can not only increase the speed and convenience of the donation process, which can significantly enhance the overall donor experience. This can ultimately lead to higher donor retention and increased fundraising success, which is crucial for nonprofit organizations that rely on donations to sustain their operations.

Furthermore, RFPs can help nonprofit organizations reduce administrative costs associated with manual payment processing methods. With RFPs, nonprofit organizations can automate payment processing, cutting down on manual work and letting employees focus on more critical tasks[2].

RFPs represent a powerful tool for nonprofit organizations and charitable institutions to request donations from their supporters[20].

By utilizing RFPs, these organizations can increase the speed and efficiency of their fundraising efforts, including non-profit fundraising for disaster relief where instant and 24x7x365 are even more important. RFPs may also enhance the donor experience, and reduce administrative costs.

Insure Disasters Are Handled

Instant payments, particularly when integrated with Request for Payment (RFP) functionality, can revolutionize the way insurance claim payouts are managed. Once an insurance claim is approved, an RFP can be generated and linked to the specific claim, allowing the insured to initiate their preferred method of payment. This not only expedites the disbursement process but also enhances the customer experience by providing more control over how and when they receive their funds. The RFP can also include remittance advice, allowing for seamless reconciliation and auditing on the insurer's end. By reducing manual processes and waiting times, instant payments with RFP functionality bring efficiency, transparency, and flexibility to insurance claim payouts, thereby improving both operational workflows for insurance companies and the financial experience for claimants.

Government agencies

Government agencies are responsible for collecting payments for various fees, fines, tolls, and other payments. In the digital age, the Request For Payment (RFP) system has emerged as a highly effective tool for government agencies to request payments[10].

RFPs can be particularly useful when individuals or businesses owe payments to government agencies. By utilizing RFPs, government agencies can send payment requests in real time, enabling them to receive payments quickly and efficiently.

RFPs can significantly improve the efficiency of government operations by automating payment processing and reducing the administrative burden associated with manual payment methods. This streamlined approach to payment collection can reduce the costs associated with traditional payment methods and enable government agencies to focus on other critical tasks.

In addition to improving government efficiency, RFPs can enhance the overall user experience by providing a convenient and easy to use payment system. Recipients can respond to payment requests quickly and efficiently, leading to higher satisfaction and engagement with government agencies.

Moreover, RFPs can increase transparency and accountability in government operations by providing a traceable payment system. This could enhance public trust and confidence in government agencies and ensure that payments are accurately recorded and processed[2].

RFPs represent a powerful tool for government agencies to collect payments. By utilizing RFPs, government agencies can improve the efficiency of their operations, reduce costs, and enhance the overall user experience.

How can faster payments benefit financial institutions?

The immediate and obvious benefit is that providing instant payments supports your account holders transactions and internal liquidity needs. That is just the beginning of the benefits, as instant payments also will provide financial institutions with opportunities to attract new and retain existing customers by offering competitive instant payment services[3]. Sooner rather than later, if you will not or do not offer instant payments, someone else will or already is.

Instant payments also allow the financial institution to grow revenue by attracting new account holders and collecting more transaction fees or generating fees from new services or new product offerings that instant payments enables[1]. Instant payments create the opportunity for financial institutions, on their own or in collaboration with Fintechs, to create new products that use instant payments.

Additional benefits also include reduced costs through increased efficiency and the creation of new products and services. When preparing for your launch of instant payments, flesh out the many advantages its service can bring to your financial institution and account holders. To get started, identify what benefits can be created and what problems can be solved with instant payments internally as well as externally. When done internally, this not only reduces risks, but also set a lead by example model that can be communicated to your customers/members as things you can now enable them to do as well.

Here are some examples of how financial institutions can use the instant payments service for internal purposes:

- Reimbursing employee expenses. Instant payments allow businesses to reimburse employees immediately, if needed[4].

- Payroll and bonuses. Say a payroll payment was missed or needs to be corrected. You may choose to stick with an Automated Clearing House (ACH) for payroll deposits and have the instant payments as a stopgap to resolve issues, such as a late entry that misses the ACH window or issue a payment that needs to be paid immediately[4].

- Incentive pay or corrections.

- Disbursing 401(k) loans or investments faster than ever before to get the money in the hands of those who need it.

- Funding of customer or member loans, such as auto loans, mortgages, home equity line of credit (HELOCs), prepaid credit or debit cards.

- Offering liquidity to other financial institutions.

That is just the beginning of the many ways instant payments can be used within a financial institution. What about making its services available to external partners, merchants, corporate or retail, and consumers?

Consumers, everyday account holders, will see some of the most significant benefits. And this creates incredible opportunities for additional services and products a financial institution (FI) can leverage for customer satisfaction and revenue growth[4]. Advantages to the consumer include instantly paying loans or other credit accounts to avoid incurring late fees, for example. It also could be the funding conduit to their healthcare accounts[4]. Consumers could have the ability to transfer funds between accounts at separate financial institutions, what we refer to as "account-to-account" or "me-to-me" transfers. This scenario opens the door to even more ways instant payments could benefit account holders[4]. Numerous opportunities are possible, from funding a brokerage account at a different financial institution to taking advantage of an investment opportunity or a hot stock tip within an online application and platform. What if you need money moved into or out of a product or service related to your investments not held at a financial institution? Think Acorns, Robin Hood, or Stash, as there are several today.

Consumers may open a new account for specific savings or other purposes. instant payments could be used for funding or cashing out a digital wallet used online or at another institution.

These examples represent the tip of the iceberg of consumer use cases, and in many of these examples, small businesses benefit as well.

Let us move along and look at the benefits and abilities that instant payments bring to a financial institution's business customers.

The use cases and benefits include using RFP capabilities, as previously mentioned when discussing consumer uses, also bears advantages for businesses[2]. Like individual account holders, businesses also can pay recurring and one-time bills (e.g., utilities, leases, suppliers, and loans). They also can pay yearly or quarterly taxes at the last minute without building in time for snail mail.

Businesses and merchants can have instant cash concentration abilities and consolidate excess cash from different accounts from various subsidiaries into a centralized account for better cash management. This functionality could redefine the sweep process, as businesses can perform and manage their sweeps by instantly transferring cash between accounts at multiple financial institutions. For example, they could move money from a non-interest-bearing cash bank account that exceeds or falls short of a predetermined level into or out of an interest-earning investment account at the close of the business day.

Businesses could use instant payments to complement their existing payroll abilities by having the ability to instantly disburse payroll or payroll exceptions (e.g., errors, incentive pay, and final paycheck). Businesses also can pay service providers and suppliers for inventory, services, and rent. They could even use instant payments to pay one-time transactions such as insurance claims or rebates instantly.

Use case: A Modern Solution to Bank Robberies

While we have explored several benefits, including increased convenience and better cash flow management, a significant benefit may take time to realize with instant payments[5]. Interestingly, instant payments have contributed to reducing the chances of bank robberies, as seen in Denmark[6].

Denmark is among the countries leading the way in adopting instant payments. In 2014, the Danish Central Bank (Danmarks Nationalbank) and the Danish banking sector introduced the Real Time 24/7 clearing system[25]. This system, also known as Straksclearing, enables Danish citizens to make instant payments between participating banks 24 hours a day, seven days a week.

Denmark has seen a remarkable decline in bank robberies since adopting instant payments. According to Danish National Police statistics, bank robberies dropped from 114 in 2009 to 12 in 2020 (source: Danish National Police, 2021)[6]. While several factors have contributed to this decline, the widespread use of instant payments has significantly deterred bank robberies.

Why is this a use case and how can this benefit financial institutions?

Reduced risk of physical robberies: With instant payments, banks, and credit unions can limit the amount of cash they keep on hand, making them less attractive targets for robbers. This reduces the potential financial losses from theft and minimizes the risk of harm to employees and customers[6].

Lower security costs: The reduced risk of robberies allows banks and credit unions to decrease security expenditures. This includes costs associated with security personnel, surveillance equipment, and insurance premiums[6]. These are savings that may be used to increase cybersecurity systems.

Instant payments have revolutionized the banking industry, providing numerous benefits to banks, credit unions, and customers. In Denmark, the adoption of instant payments has been linked to a significant decrease in bank robberies, making technology an essential tool for enhancing the safety and security of financial institutions[6]. As more countries, including the U.S., adopt instant payment systems, we can expect similar benefits in reduced crime rates and improved efficiency in the banking sector.

So many options, what to choose?

How do you determine what to offer? That is up to you and something you will want to have identified as you prepare your product and service roll out.

Some questions to ask and issues to identify that will help with knowing where to begin offering faster payments include:

- Are there specific reasons why business customers are interested in making payments instantly?

- Are business and retail customers seeking confidence that funds are available to them, without concerns of insufficient funds or delays, funds availability and merchant chargebacks?

- Are customers looking to take advantage of opportunities that an instant payment can provide which other payment channels may not support, such as sending funds immediately to a college student?

- Are customers especially interested in the transparency and confidence that an instant bill payment can provide?

- Are business customers interested in having access to the additional payment data that may be included using RFP options?

There are many ways that instant payments will be able to add to and complement existing products and services. Make sure you know how you will be offering instant payments as you prepare for launch.

Instant Payment Use Case Benefits by Industry

From offering businesses various benefits such as improved cash flow, reduced transaction costs, and enhanced customer experience, let's further explore business use cases for instant payments, with a focus on the key industries and key players in the instant payment ecosystem and the potential impact of these payments.

Instant payment systems are revolutionizing the way businesses and individuals carry out transactions. Identifying the specific needs of business and retail customers is essential to determining which faster payment offerings to provide. Instant payments offer numerous benefits across various industries and use cases, including P2P payments, bill payments, online and in-store purchases, retail, the gig economy, and B2B payments[9].

Retail Industry

One of the major sectors that has embraced instant payments is the retail industry. With the rise of e-commerce and digital transactions, instant payments have become an essential tool for retailers to offer a seamless and frictionless shopping experience to their customers[1].

Point-of-Sale (POS) Payments

Instant payments at POS terminals allow customers to make purchases without the need for physical cash or cards[30.] This not only speeds up the checkout process but also reduces the risk of fraud and chargebacks.

QR Codes

QR codes can significantly enhance the convenience and efficiency of instant payments. By simply scanning a QR code with a smartphone, consumers can quickly initiate a secure transaction without the need for manual input of account numbers or other payment details. This not only speeds up the payment process but also minimizes the risk of errors. For businesses, QR codes can be generated to include specific transaction information, facilitating automated reconciliation and streamlining Accounts Receivable. Additionally, QR codes can be used in both online and offline settings, offering a versatile and accessible means of leveraging instant payment capabilities for a broad range of use-cases.

Online Shopping

Online retailers can use instant payment methods like digital wallets, mobile payments, and other app-based solutions to reduce cart abandonment rates and improve the overall customer experience[30].

Gig Economy

The gig economy, characterized by flexible, temporary, or freelance jobs, has experienced tremendous growth in recent years[27]. Instant payments play a crucial role in meeting the needs of gig workers and their employers.

Instant payments enable gig economy platforms to pay workers immediately upon completion of a task or job, improving their cash flow and overall satisfaction[30].

By utilizing instant payments, gig economy platforms can better manage their cash flow, ensuring they have the funds to pay workers promptly and efficiently[30].

Instant Payment Use Case benefits by use case type

Business-to-Business (B2B) Payments Benefits

Instant payments are also transforming the business-to-business (B2B) payment landscape by enabling faster and more efficient transactions between businesses[1].

Improved Cash Flow Management

Instant payments allow for real-time settlement of transactions, reducing the need for businesses to maintain large working capital buffers and improving overall cash flow[1].

One of the most significant benefits of instant payments for businesses is the ability to improve cash flow management. The immediate availability of funds allows companies to better plan their financial strategies and investments, leading to more efficient utilization of their resources[1].

Streamlining B2B Payments

Instant payments have the potential to streamline B2B transactions by reducing the time it takes for payments to be processed, settled, an reconciled. This reduction in payment processing time can help businesses save time, minimize errors, and increase overall efficiency. RFP and remittance details are just some of the functionalities of instant payments that can streamline end-to-end flows.

Enhanced Customer Experience

Instant payments can significantly improve customer experience by providing immediate payment confirmation, which helps build trust and satisfaction among customers[4]. This can lead to increased loyalty and repeat business.

Reduced Operational Costs

By eliminating the need for manual payment processing and reconciliation, instant payments can help businesses reduce operational costs. This efficiency can lead to cost savings, which can be passed on to customers in the form of higher discounts, faster payment receipt, lower prices or improved services[2].

Increased Transparency and Security

Instant payment systems often include enhanced security features such as end-to-end encryption and real-time fraud detection, which can help businesses protect their transactions and customer data[9]. Additionally, these systems provide a greater level of transparency, allowing businesses to track the status of their transactions in real-time.

Person-to-Person (P2P) and Account-to-Account (A2A) Payment Benefits

P2P payments

With P2P payments, consumers can easily send money to friends or family members in need. This can be particularly useful for splitting bills, sharing expenses, or sending money to loved ones in emergency situations. Instant payments make it possible for funds to be transferred in real-time, eliminating the need to wait for days for the money to clear[4].

A2A (Me-to-Me) Payments

Instant payments offer significant advantages for both account-to-account and "me-to-me" transfers, which involve transferring funds between one's own accounts. The immediacy of instant payments ensures that funds are available almost immediately, eliminating the waiting periods often associated with traditional bank transfers.

This is particularly beneficial for individuals who need to move money quickly for time-sensitive investments, bill payments, or emergencies.

Consumer-to-Business (C2B) Payment Benefits

Bill payments

Instant payments also can be used for bill payments, allowing consumers to pay their bills quickly and easily. This can be particularly useful for individuals who are short on time or who need to make a payment immediately to avoid late fees or penalties. Many utility companies, credit card companies, and other service providers now offer instant payment options, making it easy for consumers to pay their bills on time[30].

Online purchases

Instant payments also can be used for online purchases, allowing consumers to make payments quickly and securely. With instant payments, consumers can complete their purchases in just a few clicks, without having to enter their payment information multiple times or wait for the payment to clear[30]. This can help to streamline the online shopping experience and make it more convenient for consumers.

In-store purchases

Instant payments also can be used for in-store purchases, allowing consumers to pay for goods and services quickly and easily. This can be particularly useful for individuals who prefer to use their mobile devices to make payments, as many merchants now offer mobile payment options[30], including pay-by-bank initiated by scanning a QR code. With instant payments, consumers can complete their transactions in seconds, without having to fumble with cash, cards, or dreadfully write a check while the checkout line backs up.

Instant payments facilitate improved cash flow, reduced operational costs, an enhanced customer experience, increased transparency, and heightened security. Retailers, gig economy platforms, and businesses can leverage these benefits to streamline transactions, optimize resources, and foster customer satisfaction[7]. As businesses prepare to roll out instant payment systems, it is crucial to consider these potential use cases and their overall impact on the customer experience and financial ecosystem. By adopting instant payment systems and tailoring offerings to customers' specific needs, businesses can unlock significant opportunities for growth, efficiency, and innovation[1].

UK leading the way in Faster Payments use cases

The UK has been at the forefront of adopting and implementing instant payment systems, transforming various sectors of the economy. The widespread use of instant payments spans P2P transactions, merchant payments, in-app and online purchases, government payments, payroll and employee disbursements, charitable donations, and financial management. The ease and speed of these transactions have improved cash flow, reduced costs, and enhanced convenience for both consumers and businesses. Moreover, the SEPA SCT Inst scheme has enabled seamless cross-border instant payments across Europe, fostering financial integration, economic efficiency, and competition in the European payments market[22]. Overall, the widespread adoption of instant payment systems in the UK and Europe has revolutionized how people and businesses manage their finances, streamlining transactions and promoting a more connected, efficient economy. The UK is one of the leading countries in adopting and implementing instant payment systems.

Here are some current use cases for instant payments in the UK:

Person-to-person (P2P) payments

Person-to-person (P2P) payments are among the most popular use cases for instant payments in the UK[14].

Retail payments

Merchant payments are another popular use case for instant payments in the UK[22]. Instant payments also are used for in-app and online purchases, allowing consumers to make payments quickly and securely[28].

Government payments

Instant payments are also used for government payments, allowing individuals and businesses to pay fees, fines, and taxes quickly and securely. The UK government has implemented several instant payment systems to make it easier for individuals and businesses to make these payments, including the Faster Payments Service and Bacs Direct Credit[29].

Payroll and employee payments

Instant payments also are used for payroll and employee payments, allowing businesses to pay their employees quickly and efficiently.

Charitable donations

Instant payments also are used for charitable donations, allowing individuals to donate money to their favorite charities quickly and securely. Many charities in the UK now offer instant payment options, making it easy for individuals to make donations in real time and support the causes they care about[28].

Financial management

Instant payments also have been used for financial management, allowing consumers and businesses to manage their finances more effectively. With instant payments, consumers and businesses can monitor their account balances in real time and make payments quickly and securely. This can help improve financial decision making and reduce the risk of fraud and financial mismanagement.

Instant Cross Border

The SEPA (Single Euro Payments Area) Instant Credit Transfer (SCT Inst) scheme facilitates cross-border instant payments in Europe. Launched in 2017 by the European Payments Council, this scheme enables individuals, businesses, and public administrations to make instant electronic euro payments across the 36 SEPA countries[29]. The system operates 24/7/365, allowing for real-time processing of transactions up to €100,000 within seconds. SEPA SCT Inst fosters greater financial integration, economic efficiency, and competition in the European payments market by standardizing and streamlining cross-border payments.

Use Cases from the land of the rising faster payment sun!

Asia has emerged as a leading region in adopting and implementing instant payment systems, offering valuable insights for other regions. Critical use cases for instant payments in Asia include P2P transactions, merchant payments, in-app and online purchases, government payments, payroll and employee disbursements, charitable donations, and cross-border payments. The widespread adoption of these systems has enhanced convenience, improved cash flow, and reduced transaction costs for consumers, businesses, and governments alike.

Instant payment systems such as India's UPI, Singapore's PayNow, China's CIPS, and Japan's Zengin System have transformed the financial landscape in the region, showcasing the potential of seamless cross-border transactions[26]. While the US-based instant payment systems remain domestic as of 2023, the advancements made in Asia demonstrate future global financial integration possibilities. The continued growth and development of instant payment systems in Asia will serve as a blueprint for other regions, fostering a more interconnected and efficient global economy.

Person-to-person (P2P) payments

Asia is one of the leading regions for adopting and implementing instant payment systems. What can we learn from their experiences?

Person-to-person (P2P) payments are one of Asia's most popular use cases for instant payments[21].

Retail payments

Merchant payments are another famous use case for instant payments in Asia. The use of QR codes in Asia has helped merchants displace the use of cash and cards as well as improved the consumer's purchasing experience.

Government payments

Instant payments have also been used for government payments, allowing individuals and businesses to pay fees, fines, and taxes quickly and securely. Many governments in Asia have implemented several instant payment systems to make it easier for individuals and businesses to make these types of payments, including India's Unified Payments Interface (UPI)[26] and Singapore's PayNow.

Payroll and employee payments

Instant payments also can be used for payroll and employee payments, allowing businesses to pay their employees quickly and efficiently.

Charitable donations

Instant payments also have been used for charitable donations, allowing individuals to donate money to their favorite charities quickly and securely. Many charities in Asia now offer instant payment options, making it easy for individuals to make donations in real time and support the causes they care about.

Cross-border payments

Instant payments also have been used for cross-border payments, allowing individuals and businesses to send money to other countries quickly and securely. This can be particularly useful for businesses that need to pay suppliers or partners in other countries or for individuals who need to send money to family members or friends living abroad.

Many instant payment systems in Asia now offer cross-border payment options, such as China's Cross-border Interbank Payment System (CIPS) and Japan's Zengin System[24]. While the U.S. based instant payment systems are domestic only as of 2023, the potential of what is possible has been demonstrated yet again.

What's your experience?

Instant payments have emerged as a game-changer in the financial industry, revolutionizing transactions and changing the way businesses and consumers interact. Unlike traditional payment methods, instant payments enable transactions to be completed in real-time, providing users with a convenient, secure, and hassle-free experience[1]. Let us explore how instant payments enhance the user experience and their growing impact on the financial industry.

Instant payments offer unparalleled speed and convenience, enabling users to complete real-time transactions 24/7. This eliminates the need for traditional payment methods that often involve long wait times and delayed transactions, allowing users to access and transfer funds instantly[9]. For instance, users can make payments, transfer funds, and pay bills in real time using instant payment systems, thereby significantly improving their experience.

Moreover, instant payments offer greater transparency and traceability, allowing users to track and monitor their transactions in real time. This gives users greater control over their finances, enabling them to track their spending and identify fraudulent transactions. With instant payment systems, users can receive real-time notifications and alerts when a transaction is made, giving them greater transparency and control over their finances[2].

Instant payments enable users to transact anywhere, anytime, and on any device. Unlike traditional payment methods that require users to visit a bank or a payment terminal, instant payment systems can be accessed and utilized from anywhere in the world using a mobile device or a computer[24]. This provides users the flexibility and convenience to make transactions on the go, eliminating the need for physical cash or visits to a physical location and enhancing their user experience.

Moreover, instant payments positively impact businesses, providing them with greater efficiency, cost savings, and customer satisfaction. With instant payment systems, businesses can receive payments instantly, reducing the wait times associated with traditional payment methods. This improves cash flow and liquidity for businesses, providing them with more excellent financial stability and enabling them to reinvest in their operations[2]. Furthermore, instant payments provide businesses with greater convenience and flexibility, offering customers a more convenient and hassle-free payment experience. This enhances customer satisfaction and loyalty, increasing sales and revenue for businesses[4].

Instant payments offer an enhanced user experience, providing incredible speed, convenience, security, transparency, and accessibility. Instant payment systems are transforming the financial industry, providing users with a more convenient, cost-effective, and inclusive payment method[30]. As the use of instant payments continues to grow, it is clear that they will play a significant role in shaping the future of the financial industry, enhancing the user experience, and contributing to many future successes in the U.S. electronic payments industry.

Conclusion

As the financial industry prepares to embrace the use of instant payments, it is crucial to recognize the need for comprehensive planning across various areas, departments, and personnel of a financial institution. Such planning involves collaboration and coordination with third-party providers, as well as consultation with legal counsel to determine compliance with applicable consumer protection laws and regulations. It is vital to be aware of the rules or laws that may impact the launch and use of instant payments.

Recognizing the benefits instant payments can provide to financial institutions will accelerate growth in this sector. The benefits include the ability to attract new and retain existing customers by offering competitive instant payment services, creating new revenue streams from transaction and service fees, and supporting liquidity needs. Additionally, instant payments create opportunities for financial institutions, on their own or in collaboration with Fintechs, to create new products that use the instant payments. Moreover, instant payments reduce costs through increased efficiency, provide real-time settlement in central bank funds, and reduce interbank settlement risk.

When analyzing use cases and ways instant payments will be offered, the focus should be on creating new products and services that solve problems and offer benefits to consumers and businesses. Consumers, for instance, can benefit from the ability to instantly pay loans or other credit accounts, transfer funds between accounts at separate FIs, fund brokerage accounts at different FIs, reload prepaid cards, and pay other people via instant person-to-person payments. Small businesses, on the other hand, can benefit from instant cash concentration abilities, consolidating excess cash from different accounts, instantly disbursing payroll, paying service providers and suppliers, and paying one-time transactions such as insurance claims or rebates instantly.

Before rolling out instant payments, financial institutions should assess the needs of their customers and identify specific reasons why business customers are interested in making payments instantly.

Customers may be seeking confidence that funds are available to them, without concerns of insufficient funds or delays, or looking to take advantage of opportunities that an instant payment can provide which other payment channels may not support.

As financial institutions prepare to implement instant payments, it is vital to recognize the numerous benefits that it can provide, assess the needs of customers, and identify the use cases and ways that the service will be offered[1]. By doing so, financial institutions can create new products and services that solve problems and offer benefits to consumers and businesses.

Chapter 6 - Implementation Financial Institution Specifics for Faster Payment Use Cases[1]

By Atilla Csutak

Much of what banks need to prepare for the implementation of instant payments include most of the things that they already do for other payment services. Additional considerations include those specific to the 24x7x365 instant experiences with their users, real-time fraud detection and mitigation, liquidity management, irrevocable credit push only payments, the economics of instant payments, and new functionalities such as those enabled by the rich data and new types and non-value ISO 20022 messages such as request for payment and request for information.

A key consideration for instant payment implementation is the use cases that their customers want and need plus those that will protect their customers from using alternative experiences that they may have started to use already.

This chapter looks at specific use cases that financial institutions can consider offering to their customers and what implementation considerations relate to these.

Use Case Examples to Consider

As the U.S. market transitions to faster payments, business cases for faster payments include instant person-to-person (P2P) and account-to-account (A2A) money movement, consumer and business bill/invoice payment, business disbursements, enhanced payroll support for select employee groups (EWA – Early Wage Access), and accelerated loan disbursement, just to name a few.

The Clearing House's RTP˚ and Federal Reserve Bank's FedNow˚ networks have the potential to transform the way financial transactions are conducted, providing instant/real-time and convenient fund transfers between parties. This summary takes a closer look at various use cases where faster payments can transform different sectors and industries, offering enhanced experiences and improving operational efficiencies.

- **Person-to-Person Payments (P2P) also known as Consumer-to-Consumer (C2C):** Instant payments have made P2P transactions seamless and effortless. Individuals can send money to friends, family, or colleagues instantly, eliminating the need for physical cash or delayed transfers. Whether it is splitting bills, repaying a friend, or contributing to a group activity, real-time payments provide a quick and efficient way to transfer funds. Examples include:

    ◉ Caregiver services, childcare, eldercare, and pet sitting.

    ◉ Extracurricular activities travel fees, sport fees, membership fees.

    ◉ Personal trainers, tutoring, music and other lessons

    ◉ Splitting the bill when dining out with friends.

- ◎ Sending money to family members.

- ◎ Buying goods from private sellers

- ◎ Paying for personal lawn care, snow shoveling, and other handyman contractors

- ◎ Gratuities and personal donations

- ◎ Child support and alimony payments

- ◎ Personal micro business payments (e.g. food carts/trucks, flea markets, farmer's market, etc...)

- ◎ Sublet rental payments

- **Account-to-Account (A2A) also known as me-to-me:** Instant payments have sped up and simplified personal and business cash management where funds can be transferred between multiple accounts owned by the same individual/joint owners or owned by the same business. These transfers can be done the moment these funds are needed or to maximize interest and investment opportunities. Examples include:

  - ◎ Transfers between checking and savings at different financial institutions

  - ◎ Transfers to and from brokerage investment accounts

  - ◎ Funding and defunding eWallets and prepaid cards.

  - ◎ Buying and selling crypto and other digital assets

  - ◎ Corporate account sweeps to and from central operating and/or interest-bearing accounts

  - ◎ Business payroll and payables disbursement account transfers

  - ◎ HSA/FSA account funding and reimbursements.

  - ◎ Other various business treasury management account-toaccount transfers including FX currency exchanges

- **Consumer Bill Payments (C2B):** Faster payments have simplified bill payments, making it easier for customers to pay for utility bills, credit card payments, loans, and other recurring expenses. Rather than waiting for checks to clear or relying on delayed payment processing, customers can pay their bills in real-time, ensuring timely payments and avoiding late fees or service interruptions. The use of ISO 20022's Request for Payment (RfP) for bills not only facilitates the immediate payment of these bills, but also provides a paperless and enhanced experience for both the biller and its customers. The remittance advice that can originate from the RfP, or recorded at time of payment, can flow through for billers to be able to automate accounts receivable straight-through-processing. Examples include:

  - Consumer utility and other similar recurring bill payments

  - Credit card and loan payments

  - Insurance premiums

  - Medical co-pay bills

  - A homeowner paying small businesses for yard maintenance and household-related services, such as mowing, mulching, roof repairs, tree removal, painting, or cleaning

    - ▪ By getting paid in real time, these small business service providers can reduce their operating overhead as funds become available instantly and avoid transaction fees that eat into their bottom lines, such as average credit card transaction fees between 1.5 to 3.5 percent

  - Education costs, such as paying for tuition, books, cafeteria, events, and tutoring

  - Property Management bills to pay rent and association fees

- ⊚ Other recurring rental bills (e.g. storage facilities)

- ⊚ Prepaid travel costs to cover one-time or recurring fees, such as excursions, luggage, car rental, hotels, vacation rentals, bus and train rides, tolls, etc...

- ⊚ Charitable giving to non-profit organizations

- ⊚ Subscription services payments for news, books, streaming services, gym and other similar memberships, etc...

- ⊚ Bill collections

- **Retail and E-commerce Purchases (C2B and B2B):** Faster payments have the potential to disrupt the retail and e-commerce sectors. Customers can make purchases online or in-store and complete the payment instantly, ensuring immediate authorization and confirming the transaction in real-time. This helps merchants with faster order processing, efficient inventory management, and quicker fulfillment of customer orders. Examples include:

  - ⊚ Restaurants, bars, and all food services

  - ⊚ Groceries, bakeries, and other similar retail locations

  - ⊚ Clothing, gifts, and other personal goods retail outlets

  - ⊚ Home furnishings and furniture purchases

  - ⊚ Hardware and other similar supplies and home/apartment goods

  - ⊚ General retail outlets (e.g. Walmart, Target, etc...)

  - ⊚ Auto and other similar service locations

  - ⊚ Hair salons, tanning, and other personal service locations

  - ⊚ Entertainment tickets costs, such as for sporting events, shows, concerts, museum entrance fees, and other entertainment and recreation entrance fees

⊚ Leisure rentals (e.g. boating, skiing, and other leisure sports rentals)

⊚ On demand versus prepaid travel expenses

⊚ Other retail point of sale purchases

⊚ Online marketplaces (e.g. Amazon, Shopify, eBay, etc…)

⊚ Online direct retail website purchases

⊚ A credit union disperses funds to a car dealership on behalf of a member

⊚ A Buy Now Pay Later provider funding a BNPL purchase to merchants

⊚ Cash on Delivery (COD) payments

⊚ Parking facilities and metered services

⊚ Vending machines

- **Business-to-Business Payments (B2B)**: Faster payments have transformed business-to-business (B2B) transactions by providing faster and more efficient payment solutions. Suppliers and vendors can receive payments instantly, improving their cash flow management and reducing the need for credit extensions. Faster payments also streamline payment processing and reconciliation processes, enabling businesses to reduce DSO (Days Sales Outstanding) and maintain more timely and accurate financial records. Examples include:

  ⊚ Supplier Payments

  ⊚ Farmers ordering fertilizer, feed, and other inputs

  ⊚ Home builders/remodelers purchasing construction supplies, equipment, etc...

  ⊚ Freight and other transportation service payments

  ⊚ Title Companies processing real estate closing payments and disbursements.

- ◉ Trade finance funding

- ◉ Invoice Factoring, Merchant Cash Advances, and other business loans

- ◉ Businesses purchasing postage and mailing lists

- ◉ Healthcare insurance payments to healthcare providers

- ◉ E-marketplace supplier payouts

- ◉ Merchant services settlement

- ◉ Retirement account payroll deduction transfers to 401k accounts

- **Business-to-Business Requests For Payments and e-Invoicing (B2B RfP & e-Invoicing):** The use of Request for Payment (RfP) for B2B payments provides a valuable means of paperless billing and payment initiation. RfPs can provide the essential information for business customers to initiate a more streamlined and automated accounts payable process, especially for simple B2B invoices where the ISO 20022 RfP message can contain all the information necessary to initiate these payments. For complex B2B invoicing requirements, such as those that must be matched against purchase order and receiving document line items, the RfP can accompany an e-invoice which can automate these accounts payable records where the RfP schedules the payment to be made in time for applicable early payment discounts or up to the last moment that the invoice is discount eligible or due.

- **Payroll, Gig Economy, and Freelancer Payments (B2C):** Faster payments can accelerate the gig economy. Independent contractors, freelancers, and gig workers often rely on prompt payment for their services. Faster payments allow gig workers to receive their earnings immediately after completing a job, ensuring financial stability, and eliminating the wait for traditional payment cycles and methods. Employees can get

189

paid daily, more frequently, or when something got missed during payroll processing. Employees can also get more timely expense reimbursements and payroll advances. Examples include:

- ⊚ Payroll earned wages paid daily or just more frequently
- ⊚ Employee expense reimbursements
- ⊚ Employee payroll advances
- ⊚ Contractor and subcontractor payouts
- ⊚ Uber, Lift, and other similar gig driver payouts
- ⊚ Truck driver payments
- ⊚ Food and package delivery gig workers
- ⊚ Freelance contractor payments

- **Business to Consumer Disbursements (B2C)**: Instant payments can be used for a wide variety of disbursements which can benefit from the speed, 24x7x365, and other aspects of instant payments. Examples include:

  - ⊚ Insurance company payment for a claim to the policy holder.
  - ⊚ Marketplace seller disbursements (e.g. eBay)
  - ⊚ Aibnb, VRBO, and other similar vacation rental payments to the home/apartment owners
  - ⊚ Merchant refunds to customers
  - ⊚ Merchants pay out rebates, promotions, incentives.
  - ⊚ Gaming venues paying winner disbursements.
  - ⊚ Royalties, referrals, and other commission payments.
  - ⊚ Child support payments paid out of payroll deduction

- ⊚ Non-for-profit disbursements to those they provide aid to

- ⊚ Annuity disbursements

- ⊚ Pharmaceutical and other trial participant payouts

- ⊚ Legal settlement disbursements

- ⊚ Restaurant pooled tip disbursements

- **Government To and From Consumers (G2C,C2G), Businesses (G2B, B2G), and intra-government (G2G):** Federal, State, City, County, and other government entities have numerous types of payments they pay out as well as receive which could benefit from the speed, 24x7x365, and data that enables improved transparency and reconciliation benefits of instant payments. Examples include:

  - ⊚ Tax refunds and tax payments

  - ⊚ Benefit disbursements

  - ⊚ Disaster relief disbursements (e.g., FEMA)

  - ⊚ Special government aid programs (e.g., COVID)

  - ⊚ Permits, fines, and other similar fees

  - ⊚ Intra-agency fund transfers

  - ⊚ Federal fund transfers to and from states

  - ⊚ Government grant disbursements

  - ⊚ Lottery payouts

  - ⊚ Bail bonds

- **Interbank Settlement:** Financial Institutions transfer funds between each other for a number of use cases that are currently done by ACH and wire. Instant payments can enable these to be done faster, 24x7x365, and accompany additional reconciliation data through the capabilities of ISO 20022 remittance advice details included with or sent separately and linked to the instant funds transfer message. Examples include:

  - ATM Network Settlement

  - Corporate credit union and bankers bank transfer with the financial institutions they serve

  - Any type of financial institution to financial institution funds transfer.

- **International Cross-Border Payments:** Faster payments have started to disrupt the traditional cross-border remittance market. Sending money internationally can now be done in real-time, providing individuals with faster and more cost effective options compared to traditional wire transfers. Faster international money transfers reduce the need for intermediaries, lowers costs and improves transparency. Faster payments can be used on the sending, receiving, or both sides of a cross-border payment. Examples include:

  - A faster payment can load funds to prepaid or debit card that can be used in other countries

  - A faster payment can be used on the receiving side to immediately disburse funds to a beneficiary account

  - Faster payments can be used on both sides of a cross-border payment in a scheme called Payment-versus-Payment (PVP).

    - This is where the origination instant payment that initiates the payment only concludes once the receiving institution is confirms receipt through the separate instant payment system.

Now that we have you thinking of the possible use cases, you need to decide which ones to prioritize and focus on. Hopefully this also helps you think of new use cases and learn from others that you can keep your eye and ears open to the evolutionary and revolutionary emergence of use cases you notice in the market.

Use Case Implementation Considerations

Next, we shift focus from what use cases you may want to offer to considerations of how you can implement and manage these use cases as instant payments.

- **Disruption of Legacy/Traditional Payment Methods**: Faster payments have started to disrupt traditional payment methods like checks, card, and wire transfers. With the ability to transfer funds instantly, real-time payments offer a more efficient alternative, reducing reliance on manual check processing. This disruption fosters cost savings, eliminates administrative burdens, and improves operational efficiency for both individuals and businesses.

- **Use case economics and customer demand:** As faster payments may displace legacy/traditional payment methods for some use cases, the economics of choosing which use cases to implement and how to make money on them needs to be considered. This requires that the ROI not only covers the costs of implementation of the use cases, but also accounts for operational savings and potential lost revenues and their associated costs. When looking at costs, be sure to consider all direct as well as indirect costs, including customer service, chargeback services, fraud and other risks, compliance requirements, potential losses, as well as other not so obvious indirect costs. You need to consider the costs for what might be displaced as well as what it might be going forward in these faster payment use cases, where for instance chargeback service costs are eliminated or at least reduced by instant payments.

- One might think then that the best use cases are those that have the lowest current revenue and relatively high costs, but this should not outweigh primarily understanding the wants and needs of your customers where if you don't go to market in time, you may have lost the revenue streams if these customers switch to other services that do meet these demands.

- The economics of different use cases may warrant different models and rates, such as payment submitted last minute before they are due for a bill might be worth a premium fee versus this paid early or just scheduled earlier to pay the moment before it is due.

- What you want to avoid is losing a customer over new or higher fees that do not come with an added benefit to the customer. This also requires keeping tabs on the market of what others in the industry are charging for instant payment use cases.

- Business customers that can benefit from additional capabilities and/or data that saves them on their current direct and indirect costs plus enhances reconciliation, transparency with their customers and vendors, and enables more automation that these parties may be willing to pay for. Small businesses will be especially happy to reduce their current relatively high fees for merchant services interchange, chargeback, and other fees where they can pass on the savings in for form of gaining customer loyalty versus first adding surcharges to their customers paying by card. Use cases specific to these current economics and frustrations may be those that you least want to displace, but these customers will easily switch to another service that beats you to the punch. In this scenario, the opportunity loss is a key indirect cost that you cannot ignore.

- **Use case marketing, education, and awareness:** You can't assume that if your customers are not asking about instant payments, that they don't need them. You might ask them if they want or need instant payments and they may say do you mean PayPal, Venmo, or Zelle.

What you can and should do is educate them by asking them about specific use cases, by asking them specific questions such as...

⊚ Is your employer or the gig service you are working with able to pay you as frequent as daily ore as you earn your wages?

⊚ Do you have a need for electronic remittance details on payments you receive from your business customers?

⊚ Would you like to be able to receive a request to pay a bill and be able to pay schedule it the last moment before its due?

⊚ Do you have some bills that you need to pay immediately in order to avoid late fees or being shut off from a service?

Your marketing communications should be educational and specific to the use cases that you are offering to applicable specific customers, clients, or users. It should generate awareness of experiences and benefits that they can share with peers and be able to intuitively use immediately through existing logins to your online and mobile applications.

Before you communicate with your customers, clients, or users, you need to ensure that everyone in your organization is educated and made aware of what instant payments are all about as well as any use case specific offerings you are marketing. Consider having them all read this book to get started even planning your roadmaps for faster payment journeys.

- **Use case registration, opt-ins, and opt-outs:** Some use case functionalities may require additional opt-in registration or they can be assumed to be auto registered with an opt out option. For example, registration of businesses that want to originate requests for payment bills might require registration as well as implementation setups. On the receiving side of request for payment, these users may just be able to opt-out when they get notification of their first bill pay request for payment where they may want to just opt-out of continuing to review specific bills or opt-out of receiving all bill pay requests.

Lessons should be learned from any of those that have implemented some other bank account integrated service where the user registration of the additional service caused friction even before starting to use it. The onboarding experience is just as important as the transactional user experiences. Both must be intuitive, easy and avoid having to enter information that the service should already have about the user that has originally been onboarded when their account was opened.

- **Single sign-on and use case specific additional authentications**: Users should not have to setup and use a separate login for using new instant payment use cases. They should be able to use a single sign-on from their current logins. Some use case confirmations may require additional biometric or one time passcode authentications. These requirements may not only be use case specific, but also amount and other criteria conditional such as the detection of unusual or suspicious transactions.

- **Fraudulent Transactions:** When implementing faster payment use cases, financial institutions will need to identify and mitigate fraudulent transactions that might be happening at lightning speed. The ability to identify such transactions in real time requires complex analytical capabilities and potentially the use of artificial intelligence and machine learning to be able to assess the transaction and provide predictive insights in real time.

Some use cases, such as those that involve requests for payment may require additional fraud registration procedures, controls, monitoring, prevention, and mitigation requirements to ensure that the party initiating the requests is not a bad actor. Appropriate KYC and unusual or suspicious outbound requests as well as inbound and outbound payments monitoring is required to detect such activities in real-time as well as over various time and date windows.

Consider applicable limits and other controls and procedures with the assumption that bad actors will attempt to defraud your customer, clients, and users into paying for fraudulent requests for payment either originated through the instant payment rails or initiated through scam techniques.

Since the US instant payment rails are only credit-push payments, unauthorized debit pull payments are not a fraudulent transaction type that you need to worry about. Instead, you need to still consider prevention, detection, and mitigation of authorized push payments.

The FedNow instant payment service requires that any fraudulent transactions get recorded and notified using one of the types of fraud as determined by its Fraud Classifier model.

FedNow also has a negative list feature which enables participants to block specific accounts from sending and/or receiving transactions from these blacklisted accounts which may have been determined as suspicious or involved in a fraudulent transaction.

- **Use case specific limits:** The instant payment networks have their own transaction limits, and they allow their participating financial institutions to set their own limits within the network caps. These limits can be applied across all use cases or you can more logically consider different limits for different use cases.

Different use cases may require different limits, such as high volume-low value point of sale transactions at restaurants, bars, groceries, and other similar retail transactions would have very different limits than low volume-high value use cases such as for real estate closings, car, boat, furniture, and other major purchases. Some financial institution implementations may allow their users to set their own limits within their caps which may or may not be use case specific. These are important considerations in the controls and user functionalities that can help not only prevent or limit fraud, but also prevent end users entering extra zeros by mistake when initiating a payment.

- **User preferences and controls:** This is where financial institutions and payment service specific implementations can provide some differentiation and special features for their users. They can enable users to set their own preferences and controls such as...

  - Transaction limits for all instant payments

  - Transaction amounts that require extra confirmation or authentication

  - Secondary approval amounts that require joint account holders or additional business account users to confirm these transactions

  - Automatic approval of certain request for payment use cases to pay these immediately upon receipt of the RfP or automatically on an applicable due date

  - All of the above with optional use case specific controls

  - Optional controls and limits for specific billers, merchants, or other parties that you can set for approving RfPs or initiate sending instant payments to

- **Payment Message Use Case Applicable Content:** When financial institutions consider implementing instant payment use cases, they need to adhere to the ISO 20022 standards and the corresponding instant payment platform implementation specific rules of the ISO 20022 standard. Specific considerations should be considered for different use cases where different types and levels of information may need to be included in the payment message or sent separately and linked to the payment message. The ISO 20022 messages have hundreds of types and levels of information that the message can include. Being consistent is an essential factor in the long-term success of the use of these messages. Some use cases may want and need to include structured remittance details, such as those for B2B transactions in order to enable accounts receivable straight-through-processing. Other use cases might want to provide line-item level information and additional information about a payment involving a purchase of goods of services. Industry collaboration use case specific guidelines and best practices have started to emerge, such as the FedNow Communities' document specific to RfP for Consumer Bill Payment and the X9 ISO 20022 Market Practices Forum's paper on best practices for the ISO 20022 payment remittance advice.

Faster Payment Implementation Success Factors

Faster payments refer to an electronic payment system that enables immediate fund transfers between accounts, offering convenience, speed, and enhanced financial transactions. Let's next look at key factors that contributed to the success of financial institutions that have already implemented instant payments and the benefits realized for itself and its customer base.

Banks recognized the changing needs of their customers and the evolving payments landscape. Traditional payment methods, such as checks, cards, ACH, and wire transfers often involved delays and inefficiencies. Financial institutions rose to the challenge to meet the needs of the marketplace and consumer demand. They did so by implementing faster payments services that provide customers with instant and seamless payment capabilities.

The implementation processes included:

- **Technology Infrastructure and Integration**: Financial institutions needed to invest a significant amount of money to upgrade their existing technology infrastructure to support a 24/7/365 payment rail. This included implementing modern payment processing systems (hubs) and integrating them with the existing infrastructure (core system). Financial institutions had to collaborate with vendors who specialized in real-time payment solutions to ensure a smooth and efficient integration process.

- **Regulatory Compliance**: Financial institutions needed to closely collaborate with regulatory bodies and industry organizations to ensure compliance with applicable laws and regulations. This involved understanding and adhering to regulatory frameworks governing instant payments, such as those related to risk management, anti-money laundering (AML), and data security.

- **Industry Collaboration**: To successfully implement real-time payments, financial institutions worked in partnership with other financial institutions that already had real-time payments implemented, as well as with payment processors and other industry partners. This collaboration was crucial to establish interoperability and connectivity between different payment systems, enabling seamless transactions across various banking platforms.

- **Solution Scalability**: Keep in mind that every faster payment rail introduced globally started with low volumes. Gradually, these volumes increased beyond expectations. Along the way scalability performance assessments were used to adjust for increased volumes and any blips experienced along the way

- **Channel Agnostic**: While implementing a faster payment rail, financial institutions must remain channel agnostic to provide a consistent experience to all their customers.

- **Customer (internal and external) Education and Support**: Recognizing the importance of customer education, financial institutions needed to develop comprehensive educational materials and launch awareness campaigns to inform customers about the benefits and functionalities of real-time payments. Financial institutions also needed to expand their customer support channels to address any inquiries or issues related to real-time payments in a 24/7/365 environment.

- **Use Case focuses**: The focus of implementing specific use cases based on their customer demands and optimal ROIs for the financial institutions and their customers were key success factors that could be more easily identified with the benefits of each of the use cases implemented.

The benefits and outcomes realized included:

- **Enhanced Customer Experience:** The implementation of faster payments can significantly improve the customer experience at financial institutions. Customers could initiate and receive payments instantly, eliminating the delays associated with traditional payment methods. Faster payment capabilities offered convenience, reliability, and efficiency, resulting in higher customer satisfaction and loyalty. Finally, financial institutions identified ways to manage customer complaints and queries to avoid regulatory issues. Regulatory compliance did lead to some initial additional expenses.

- **Increased Transaction Efficiency:** Faster payments streamlined transaction process for financial institutions and their customers. The immediate transfer of funds reduced administrative overhead, eliminate manual processing, and improved cash flow management. This increased transactional efficiency resulting in cost savings for financial institutions and quicker access to funds for their customers.

- **Competitive Advantage:** By offering faster payment capabilities, financial institutions achieved a competitive advantage in the market. They positioned themselves as forward-thinking and client-driven, catering to the evolving needs of customers in an increasingly digital and fast-paced financial ecosystem. The implementation of faster payments attracted new customers and helped retain existing ones.

- **Business Expansion:** The successful implementation of faster payments opened doors for financial institutions to explore new business opportunities. By leveraging the faster payment infrastructure, financial institutions ventured into partnerships with Fintech companies, e-commerce platforms, and other industry players. This accelerated the expansion of their services and revenue streams. Success of faster payments depended on the kind and variety of use case specific benefits that customers could easily understand and realize with the assistance of their financial institution.

- **Improved Risk Management**: Instant payments provided financial institutions with better risk management capabilities. The instant transfer and verification of funds reduced the risk of fraudulent transactions, insufficient funds, and payment reversals. This provided enhanced security and built trust with its customers.

Instant Payment Implementation Challenges

One of the biggest roadblocks is the intricate and archaic legacy core banking systems. They are often decades old, mainframe based platforms that support a bank's back-end operations across core functions such as account opening, account set up, transaction processing, deposits processing, loan processing, and more. By and large, legacy core platforms are incompatible with real-time payment infrastructures and have their own regulatory compliance standards. In addition, they do not scale well and have their own security risks and other challenges associated with performance that are not equipped to provide secure around-the clock instant payment services 365 days a year.

As legacy business and technology professionals are retiring in droves, they are leaving behind a big vacuum of technical and institutional knowledge. For example, COBOL programmers are aging out of the workforce, making them increasingly scarce and expensive to replace. So too are long-tenured professionals with deep knowledge of a bank's existing systems (and the embedded business rules that enhance the bank's competitive advantage). What is more, universities and colleges are no longer teaching COBOL language-based programming classes. And yet COBOL language-based programs are still widely used in applications supporting mainframe computers, such as large-scale batch and transaction processing jobs. And it does not help that today's top IT talent has little desire to work with outdated systems and platforms, making it even harder for financial institutions to attract the experts and innovators necessary to survive and thrive in the age of digital transformation.

As financial institutions move from a customized set of legacy mainframe applications to a configurable set of cloud-native applications, it is essential to preserve the valuable intellectual property in the existing business rules. In-depth, accurate knowledge of business rules can help organizations resolve critical issues that might otherwise impede their efforts at application modernization and digital transformation. Also, the right modernization approach can build on this knowledge in ways that preserve hidden intellectual property for use in upcoming projects, helping financial institutions generate the maximum business value from existing and future applications.

Financial institutions need to be aware of other challenges littering the road to incorporating instant payments into their existing systems and services. Be prepared to work with a diversified lineup of vendors and know that each one may provide similar specialized services tailored to meet specific requirements. However, integrating these vendor services into a financial institution's architecture on the back end is not a walk in the park. Collaboration with multiple vendors is still an overly complex commitment, as financial institutions are left with the burden of continuous development, support, and maintenance costs of these disparate systems.

Considering the mix of challenges facing the transition to faster payments, it is important to tackle the climb up the mountain with a coherent strategy. That process should include assessing your current legacy core platform, identifying gaps and requirements, selecting the right platform (payment hub) solution, and integrating with and managing data migration and system interfaces. Next, consider the following as well:

- Collaborating with IT and business units

- Applying an incremental implementation approach

- Ensuring data integrity and security

- Planning for appropriate change management

- Executing disciplined project management

Typically, product management teams are tasked with providing an in-depth case study analysis that provides an overview for the financial institution's decision-makers to consider. It examines the impact that faster payment rails would have on banks with legacy core platforms. This case study empowers decision-makers to make informed choices in their journey to digital transformation. However, most financial institutions struggle to maintain a dedicated product management team and the responsibility falls to their Fintech vendors. In most instances, external vendors are not primed to meet this obligation. The bottom line for financial institutions is to adopt the faster payment platforms that make the most sense for their organization.

Here are several strategic questions to consider:

- Which real-time payments solution works best for your organization?

- What are the key challenges your organization will face while adopting real-time payments?

- How would financial institutions address migration to faster payment solutions via acquisitions, partnership with FinTechs, or vendor white-label solutions?

- While implementing a faster payment rail, how do you ensure that your solution remains channel agnostic and can provide a consistent and enhanced experience to your customers?

- How would financial institutions customize current product offerings to align with faster payment solutions?

- Do you incentivize your treasury management sales organization to sell faster payments over legacy payments?

- Are you able to identify potential fraudulent transactions in real-time that require complex analytical capabilities and use of artificial intelligence and machine learning to be able to assess the transaction and provide predictive insights in real-time?

- Last but not the least, are you prepared to identify and manage customer complaints and queries to avoid regulatory issues that could lead to additional expenses to adjust the system and achieve compliance?

By understanding the challenges, opportunities, and strategies associated with this transformation, decision-makers can make informed choices and navigate their digital transformation journey more effectively. It is crucial to recognize that each financial institution's circumstances may vary, and consulting with subject matter experts and technology providers is recommended to tailor the implementation approach to specific requirements and objectives.

Conclusion

As the financial industry evolves in the 21st century economy, the demand for instant payment capabilities has increased by leaps and bounds. By late 2017, the U.S. payment ecosystem welcomed two new entrants, Zelle and RTP. The RTP network launched by The Clearing House and Zelle via the Early Warning Services is generally available to consumers and businesses provided by their financial institutions. The journey to faster payments is still getting into full swing, but the industry can learn from the experiences that the early adopter financial institutions experienced when they launched their instant payment services. Today a majority of financial institutions are on the same 'war path.' It is a battle of the banks as they try to modernize their payment platforms so they can compete in the digital economy with functional, secure, and convenient digital payment systems for their customers. Faster payments are helping to change the economy, driving innovation and disruption by enabling new use cases, business models, and revenue streams.

The successful implementation of faster payment use cases can highlight the transformative effect of embracing modern payment technologies. By investing in technology infrastructure, ensuring regulatory compliance, fostering industry collaboration, and prioritizing customer education, financial institutions can achieve significant benefits. Faster payments can improve the customer experience, improve transactional efficiency, provide a competitive advantage, facilitate business expansion, and enhance risk management. This demonstrates the importance of staying ahead of industry trends and leveraging innovative payment solutions to meet customer expectations and drive success in the financial sector.

Focus on implementing the use cases that your clients, customers, and users want, while also taking into account all of the considerations, factors for success, and challenges.

Chapter 7 - Risk, Fraud, & Compliance

By Elspeth Bloodgood

Three things are true about risk in payments.

- All payment systems have inherent risk.

- Risk changes over time.

- Although we can predict some risk involved in new payment systems, risks make themselves clear over time and through data analysis.

Consider faster payments. From the moment the concept emerged in the U.S. banking industry, it was reflexively associated with the quip "faster payments means faster fraud." Somewhat empty of meaning on its own, the phrase was sometimes used as shorthand for the idea that to "[speed] up processing times forces systems to detect suspected fraud faster."[1] Other times the phrase was used to imply that overall fraud rates or specific new types of fraud would increase as a direct result of the new systems.[2]

Two outcomes emerged from this thinking. Financial institutions have used it as an excuse not to participate in new payment networks. And software vendors have tried to leverage it to sell updated fraud mitigation products.

It's probably more helpful to view the "faster payments means faster fraud" "truism" with a little more depth and complexity, to tease out what we are really seeing in the market, and what we might expect to come in the future.

Let's start by talking about risk from a regulatory perspective and work our way out from there.

According to the *Comptroller's Handbook*, published by the Office of the Comptroller of the Currency, there are multiple forms of risk for bank supervision purposes: operational, compliance, strategic, credit, interest rate, liquidity, price, and reputation.[3] Fraud risk itself arises out of and is an intrinsic part of these other categories. We'll review some risks below in the context of faster payments, see what we've learned so far, and anticipate where the industry is trending.

Operational Risk

"Operational risk is the risk to current or projected financial condition and resilience arising from inadequate or failed internal processes or systems, human errors or misconduct, or adverse external events."[4]

Where is the operational risk in Faster Payments? At the front door. In the Night. And in the need for human review. According to Bain & Company, "[t]he key to effective operational risk management is training people to anticipate what could go wrong especially when a business unit is about to do something new."[5]

Historically, the front door to an institution was at the branch. It could be locked and have security patrols drive by and maintain visibility to deter crime. That began to change with the adoption of digital banking, and the adoption of person to person (P2P) payments like Zelle® in 2017. Then, three years later, COVID-19 propelled the industry forward. The growth rate of mobile P2P payments in 2020 tripled what was forecast. [6] Some analysts suggest the pandemic moved companies forward several years in 2020 alone.[7] Now, institutions are looking at digital onboarding, digital card issuance, and increased support for digital payments for a new generation of account holders who are looking for banking services without ever stepping foot in the branch.

US Peer-to-Peer (P2P) Mobile Payment Users and Penetration, 2019-2026

millions, % change, and % of smartphone users

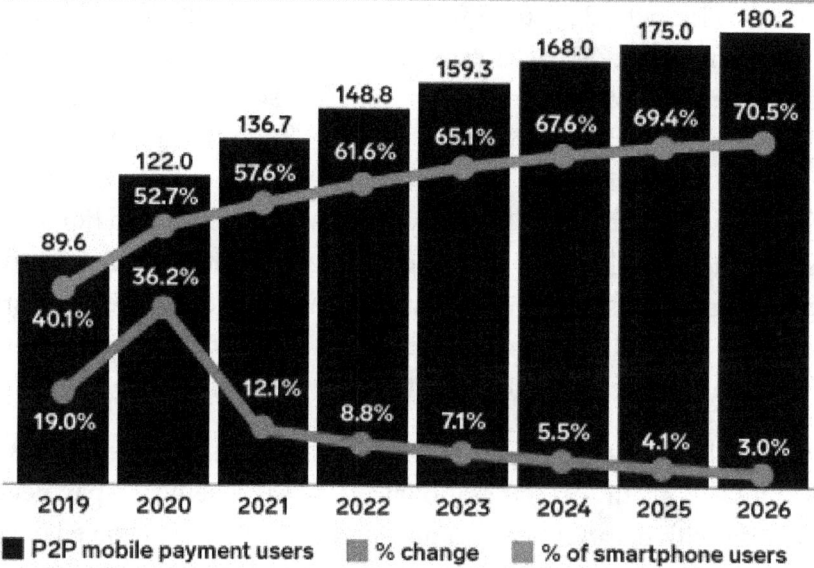

Year	P2P mobile payment users	% change	% of smartphone users
2019	89.6	40.1%	19.0%
2020	122.0	52.7%	36.2%
2021	136.7	57.6%	12.1%
2022	148.8	61.6%	8.8%
2023	159.3	65.1%	7.1%
2024	168.0	67.6%	5.5%
2025	175.0	69.4%	4.1%
2026	180.2	70.5%	3.0%

■ P2P mobile payment users　■ % change　■ % of smartphone users

Note: ages 14+; mobile phone users who have made at least one P2P transaction via a mobile phone in the past month
Source: Insider Intelligence, March 2022

274935　　　　　　　　　　　　　　　　　InsiderIntelligence.com

8

That move to digital has implications for financial institutions and their operational policies and procedures, as they look to mitigate changing risk profiles. As institutions begin to adopt faster payment networks, it is imperative that they begin to build institutional muscle sufficient to deal with new factors including:

- Know Your Customer (KYC) review for accounts opened online.

- Customer segmentation and risk analysis in light of real-time sending of payments, and non-value messages like requests for payment

These assessments help institutions identify material operational risks that could potentially go on to be significant influencers of operational losses. Material risks so identified are used in scenario analysis to estimate forward-looking events with low likelihood but that are plausible with high severity and impact.

Other operational challenges will hit institutions when the branch doors are locked. Real-time payments flow at night, on weekends and on holidays. That means institutions also have to review liquidity management procedures. Batch based infrastructure and siloed systems combine to impair visibility into real-time positions, requiring institutions to build out strategies that ensure cash is where it needs to be. Real time payments add to institution liquidity management challenges by making it more difficult to predict future liquidity positions. For institutions without this expertise, we can expect a new reliance on Bankers Banks and Corporate Credit Unions to provide settlement solutions.

What is difficult with the new payment networks is that the industry has cut its teeth on Operational Risk Management, payment networks with decades of data, and rulemaking that has evolved over time to mitigate risk. The 2023 National Automated Clearing House Association (NACHA) rule book is 716 pages. In comparison, the FedNow℠ Service, a network with no history at the time of this writing, Operating Circular 8 is only 43 pages. The RTP™ Network Operating Rules governing the network launched in 2017 are scarcely longer at 69 pages. While the weight of a rule book is not a great analog for its meaningfulness, it does indicate there are balances to be struck with industry stakeholders that haven't been captured.

This doesn't mean that we can't learn from current payment networks, threat landscape reviews, and early regulatory activity. However, it would be a mistake to assume the new real-time networks are mere variations on ACH or Wires. There are critical differences which must be taken into consideration as the industry moves forward with these new rails.

Financial institutions planning on adding new networks should move deliberately, ask questions of their vendors and processors and be ready to make program adjustments that take new learnings into consideration.

Compliance Risk

"Compliance risk is the risk to current or projected financial condition and resilience arising from violations of laws or regulations, or from nonconformance with prescribed practices, internal bank policies and procedures, or ethical standards."[9]

What we are really saying here is that, when it comes to compliance risk and with new networks, regulation follows payment activity. It does not precede it. The industry is seeing rule and regulatory changes in real time for both consumer and business transactions, across networks.

This is evident in the evolving posture of the Zelle® Network, the digital payments network run by private financial services company Early Warning Services, in turn jointly owned by Bank of America, Truist, Capital One, JPMorgan Chase, PNC Bank, U.S. Bank, and Wells Fargo. While Zelle® is an overlay service that settles primarily over the ACH and debit rails, three owner banks are settling transactions between themselves using the RTP™ network: US Bank, PNC, and BofA. The network itself has proven to be the canary in the coal mine, metaphorically serving as an early indicator of danger in systems that open consumers up to fraud in ways that are not covered under current Regulation E (Reg E) interpretation.

The Federal Reserve Board outlines rules and procedures for electronic funds transfers (EFTs) in Reg E. The regulation is meant to protect banking customers who use electronic methods to transfer money. The Consumer Financial Protection Bureau (CFPB) is an agency of the United States government responsible for consumer protection in the financial sector. CFPB's jurisdiction includes banks and credit unions. Reg E defines an unauthorized EFT as an EFT "from a consumer's account initiated by a person other than the consumer without actual authority to initiate the transfer and from which the consumer receives no benefit."[10] In FAQs issued 12/13/2021, the CFPB shared that a "credit-push P2P transfer is considered an EFT even if the payment was initiated by a third party that fraudulently obtained access to consumer's account, such as by using login credentials stolen in a data breach or obtained through fraudulent inducement. In these cases, the credit-push P2P transfer would be considered an unauthorized EFT."

This interpretation left institutions liable for transactions that resulted from an Account Takeover, even if the consumer shared their credentials. It does not, however, require financial institutions to reimburse consumers in the event they authorize the transaction themselves, even if it's for a fraudulent purpose like puppy scams, romance scams, utility scams, or the like.

However, some of the largest banks in the country were called to Capitol Hill in October of 2022 and accused of "'defrauding' customers out of at least half a billion dollars." Consumers lost about $440 million to credit push fraud on $490 billion in Zelle® payments made in 2021, according to Early Warning Services, the owner and operator of the Zelle® Network.[11]

Credit Push Fraud

Senator Elizabeth Warren subsequently released a report called "Facilitating Fraud: How Consumers Defrauded on Zelle are Left High and Dry by the Banks that Created It." In the report, Warren said "[b]ig banks own and profit from Zelle, but are failing to make their customers whole for both authorized and unauthorized fraudulent activity on the platform, despite their claims that it is safe and that they have a 'zero liability' policy for fraud."[12]

Senators called for another reinterpretation of Reg E to open the definition of fraudulent inducement to make consumers whole in cases where the payments were actually authorized.

In light of this activity, the Zelle Network quickly brought new rulemaking to the network by the end of the 2022 calendar year, with an effective date of June 30, 2023 that carved out what the network calls imposter fraud for consumer reimbursement. Payments made under the imposter fraud criteria must now be refunded to the consumer, with the liability shifted to the receiving financial institution banking the bad actor.

This is an unprecedented shift in liability, and one the industry may not be quite prepared for.

The promise of real-time payments has been with the specific characteristics of the transactions:

- Good funds

- Credit push only

- Settlement at the same time (or close to the same time) as the transaction

- 24/7/365

- Liability for sent payments falling to the sending institution

To date, there has been little, if any, conversation in the industry about liability for the receiver side of a transaction for established payment rails. That is changing quickly. In addition to the evolution of regulation and rulemaking to include incentives for receiving institutions to the real-time payment networks to establish better KYC and transaction review policies, NACHA, the rulemaking authority for the ACH Network, is also dipping its toe into uncharted waters.

NACHA released a request for comment May 3, 2023 related to credit push fraud. This followed a September 2022 release of the NACHA report *"A New Risk Management Framework for the Era of CreditPush Fraud."* [13] Nacha's previous risk management strategies for the ACH Network focused on protecting consumers, organizations, and their account-holding financial institutions from fraud due to unauthorized debits that pull money from their accounts. [14]

The new focus covers credit push fraud scenarios beyond consumer P2P, including:

- Business email compromise -- when the legitimate email account of a business officer is either compromised or impersonated and used to order or request the transfer of funds.

- Vendor impersonation fraud -- when a business, public sector agency or organization receives an unsolicited request, purportedly from a valid contractor, to update the payment information for that contractor

- Payroll impersonation fraud -— when the fraudster impersonates an employee, contacting the HR department directly or through the employer's payroll portal using stolen credentials to change the account where funds are sent

- Account takeover fraud -- when a fraudster obtains the credentials of a consumer or a business bank account and pushes credits to their own accounts.

The objectives of the new *Risk Management Framework* are to:

- Increase awareness of fraud schemes that utilize credit-push payments

- Reduce the incidence of successful fraud attempts

- Improve the recovery of funds after fraud has occurred

The request for comment contemplates rulemaking that would take effect in calendar year 2024.

How will this impact RTP™ Network and the FedNow℠ Service? It is too early to tell, but NACHA, the FedNow℠ Service, and the RTP™ Network all appear to be converging on the use of the Federal Reserve Fraud Classifier model to report scams and fraud.

The Fraud Classifier model was released in June 2020. It was developed by the Fraud Definitions Work Group, led by the Federal Reserve. The model enables payment stakeholders to classify fraud and scams in a simple and similar manner across payment types. It covers authorized party and unauthorized party fraud and is designed to enhance understanding of fraud trends, which should lead to improved fraud mitigation and easier collaboration for data sharing across the payments industry.

More recently, to help address the growing and costly problem of scams, the Federal Reserve has established a scams definition and classification work group. This group will leverage the expertise of payments and fraud experts to promote a common understanding of scams and consistent classification - extending the ability of stakeholders to report suspected scam activity in a consistent manner.

It is not unreasonable to conclude that the real-time payment networks will follow the established trend of rulemaking around credit push fraud if and when the fraud volume increases on the FedNow℠ Service and RTP™ Network. Instant requests for payment (RfPs) add a new type of potential fraud and risk by the financial institutions that enable their customers to initiate RfPs having to have a credit push payment sent back. P2P RfPs should learn from what Zelle has already in its P2P service

Looking internationally, this evolution parallels what has been happening in the United Kingdom. On September 29, 2022, the

Payment System Regulator (PSR) proposed that UK banks be required to reimburse customers within two days if they have been tricked into sending money in what is called Authorized Push Payment fraud or APP.[15] This follows a voluntary program started in 2019 in which participating banks and Payment Service Providers(PSPs) who agreed to participate would refund fraud payments. A bill working its way through Parliament with an enforcement date at the end of 2023 would replace this voluntary system with firm requirements under the regulatory guidance of the PSR. It appears that consumer advocates in the UK, like those in the United States heavily favor regulation that makes whole consumers victimized by this kind of fraud.

As much as we can learn from other systems, we also need to be aware that specific circumstances may evolve that are unique to the United States. After all, Brazil found its PIX system being used during nighttime hours to fund kidnappings. Consequently, the Pix system instituted a $200 transfer limit between individuals between 8 p.m. and 6 a.m., as well as a minimum wait time to increase transfer limits, and different transaction limits at night when kidnappings are more common. While the U.S. might not expect a similar spike in kidnappings, and is unlikely to enforce different network limits overnight, there are undoubtedly areas of risk that will evolve unique to this landscape which will need to be mitigated.

For example, in the context of the U.S. instant payment networks, stakeholders share concerns about the liabilities and risks associated with the origination of Requests for Payment. In late 2022, the RTP™ network released a rules interpretation covering the warranties of a participating institution regarding requests for payment.[16]

Under RTP™ Operating Rule VII.B.2 (effective January 1, 2023) an RTP™ Participant that submits a Request for Payment (RFP) to the RTP™ System must: "warrant to TCH and the Message Receiving Participant that the Request for Payment (1) is made for a legitimate purpose and (2) is not part of a fraudulent scheme to induce a payment; harassing, or otherwise unlawful including violations of the prohibition on unfair, deceptive, or abusive acts or practices as set forth in Title X of the Dodd-Frank Act or violations of the prohibition on unfair or deceptive acts or practices in or affecting commerce as set forth in Title 5 of the Federal Trade Commission Act;"

This rule anticipates potential misuse of a request for payment for multiple misleading or deceptive purposes. The FedNow[SM] Service has also issued an update to their operating procedures in September, 2023 to clarify the responsibilities of the Sender to monitor Request for Payment Activity.[17] This includes monitoring use of RFP to identify abuses by themselves and/or their customers. Under the guidance, monitoring must include risk-based procedures to track the volume of RFPs sent by customers to identify potential anomalous activity. In addition, Sender FIs must review, and incorporate into their monitoring, any reports or other information received regarding a customer's use of RFP. Sender FIs must also have procedures in place to investigate anomalous activity related to RFP use by itself or its customers, and following investigation must have operational procedures in place for remedial action.

Institutions contemplating allowing their business customers to originate these requests need to be prepared to carefully assess the risk associated with the origination of Requests for Payment as the networks and industry assess and offer guidance on use cases.

Institutions will also want to closely review their Know Your Customer (KYC) policies as they determine which account holders will be allowed to send payments and requests for payment on the real-time networks. Institutions will want to review new rules from the RTP[TM] Network around the types of account holders who are allowed to send Requests for Payment and set up appropriate operational procedures to review that activity on an ongoing basis. KYC should not just be limited to when onboarding new customers.

Things change and your KYC information about your customers may shed new light on their risk and qualifications for certain services.

Institutions contemplating compliance obligations should also be aware of specific controls available at the network level for the FedNow^SM Service.

The FedNow^SM Service launched in July of 2023 with the ability to set a lower transaction limit at the RTN level for participating institutions. This per transaction limit would ensure all processors are limited to a lower maximum amount than supported by the Service. A second control is the ability for an institution to upload a negative list. This list allows an institution to restrict, based on RTN and account number, send, receive or both, enforced at the network level. To take advantage of either of these controls, the institution must put in place policies and procedures for review, updating, and maintenance.

Zelle® rule changes, NACHA pending rules changes, and CFPB activity aside, the new networks are also driving potential changes in other compliance areas.

For example, to allow time for the screening of inbound transactions, FedNow^SM, RTP™ and Zelle® all have functionality that allows for "accept without posting." Accept without posting allows an institution to delay availability of funds to the recipient. It's essentially a carve out for FedNow^SM transactions of Regulation J (Reg J) subpart C. Under Reg J the receiving institution is obligated to make funds available immediately after acceptance. For FedNow^SM and RTP™ exceptions are made if there is a reasonable cause for the financial institution to believe that the user shouldn't be given availability. Reasonable cause can include sanctions, OFAC or potential fraud screening hits. One of the changes NACHA is proposing above is a similar carve out for ACH fraud transactions.

Speaking of the Office of Foreign Asset Control (OFAC), financial institutions will need to keep up with evolving guidance from this financial intelligence and enforcement agency under the U.S. Treasury Department. OFAC administers and enforces economic and trade sanctions. Automated review of payment transactions against its sanctions list is considered best practice. Both the RTP™ Network and FedNow℠ Service require participating institutions to have a written OFAC compliance program reasonably designed to promote and monitor compliance with OFAC sanctions programs and regulations.

Screening of real-time transactions gets tricky because there's not generally enough time during a transaction to perform an OFAC check and meet network Service Level Agreements (SLAs) and customer expectations for immediate funds availability. For this reason, some providers do not offer sanctions screening at the time of the transaction. Institutions should be aware of the screening capabilities of any processor's that support their instant payments network programs and monitor guidance for any changes.

As an example, guidance from 1997 related to ACH processing allowed the sending financial institution to rely on the screening of the receiving institution for timing reasons.[18] The latest guidance from September 2022 highlights the need for emerging payment technology companies to assess their sanctions risks and implement compliance programs that mitigate those risks. The new guidance doesn't specifically address the posture of institutions that screen their own account holders and rely on the other participant to the transaction to do the same but holds open the possibility this approach may need to be reviewed over time to make sure it continues to adequately address the risks of allowing payments that should be prevented from traversing the system.

Finally, institutions should review the Federal Financial Institutions Examination Council (FFIEC) guidance[19] issued in August of 2021 covering effective authentication and access risk management principles and practices for customers, employees, and third parties accessing digital banking services and information systems. By incorporating best practices, institutions have an opportunity to reduce their risk of liability for corporate real-time transactions by providing appropriate disclosures to their business customers.

Strategic Risk

"Strategic risk is the risk to current or projected financial condition and resilience arising from adverse business decisions, poor implementation of business decisions, or lack of responsiveness to changes in the banking industry and operating environment."[20]

Strategic risk for the new payment networks revolves around product strategy and timing. Institutions that go to market with the wrong vendor(s), the wrong products, without the right operational controls and institutions stand to lose more than they gain. By the same token, institutions that don't have a payments strategy or are betting the new networks won't have a meaningful impact on their long-term profitability aren't doing themselves any favors.

For many reasons, real-time payments in the United States have gotten off to a slower start than initially expected. Internationally, network mandates, limited numbers of institutions, and lack of national directories have been factors in rapid growth in countries from Brazil to Australia and Singapore. In the United States there are no mandates, and thousands of institutions. It takes time and investment for processors, core banking systems and institutions to build out, test, and begin bringing use cases to market. With the launch of the FedNow[SM] Service in July 2023, it's time to get off the fence and take action – whether an institution is a fast follower or prefers a more reactive posture.

A recent survey of Jack Henry clients found that 75% of respondents believe payments-focused Fintechs like Square, PayPal, and Venmo are their biggest competitors. Financial Institution account holders no longer bother to ask for functionality they want. They can find it elsewhere. When they go, and they will, their deposits will go with them. For example, Starbucks is holding $1.2 Billion dollars – money that hasn't been spent on coffee or treats. That's enough to establish a midsize bank. Meanwhile, fintechs like PayPal, Venmo, and Square (among others) are happy to provide expanding financial services to financial institution customers. Other, non-bank companies like Walmart, Amazon, Facebook, Apple, X, and other non-bank major organizations with huge user bases are also threats to bank customer deposits and transactions unless they sponsor or partner banks with these companies.

Institutions don't have the luxury of ignoring these threats. Jack Henry suggests they consider the following questions as they create a payments strategy[21]:

1. How are you going to aggressively and successfully compete in the near-term and long-term?

2. Do you have a payments data strategy?

3. What's your plan to comply with ever-increasing regulatory requirements?

4. Is Banking-as-a-Service (Baas) your friend or foe?

5. Are you open to open banking?

6. Are you supporting the incredible demand for real-time payments?

7. Is small-to-medium (SMB) banking important to your performance and growth?

8. Do you have a near and long-term digital currency plan?

9. Are you confident in your payments fraud strategy and solutions?

10. Is your marketing budget sufficient and are the results worth the investment?

11. Are your payments channels future-ready?

When thinking through a strategy, Jack Henry suggests a team approach with an executive sponsor to drive conversation and make sure the work doesn't end up as a document on a shelf.

Reputation Risk

"Reputation risk is the risk to current or projected financial condition and resilience arising from negative public opinion."[22]

Mention reputation risk in conjunction with real-time payments, and the topic that immediately comes up is payments fraud.

There is a reason for this. Traditional fraud prevention has largely been human-centric. Tools have been rules-based, payment method focused, heavily siloed, and typically performed on a batch or file basis. After running reports, or populating queues, fraud teams reviewed and made decisions based on the data available to them according to operational deadlines.

This is no longer enough. The obvious reason is there isn't time in a real-time payment window for human review. But there is more going on to it than that.

First, there is the industrialization of fraud. Bad actors aren't working alone. They are setting up sweatshops[23] and call centers.[24] Internationally, labor is so inexpensive that fraudsters can conduct sophisticated scams using scripts and monthly quotas in addition to bots. They evolve their attacks over time, to move seamlessly from Social Security scams to outstanding arrest warrants. New fraud vectors are likely to leverage technologies like VALL-E from Microsoft, which promises to simulate a person's voice when given a three-second audio sample.[25] It's easy to imagine misuse including spoofing spoken passwords or impersonating specific speakers for the purpose of deception.

Fraudsters using various techniques to steal account number and account access information for account takeover over the phone verbally is called vishing; through email it's called phishing; and via text message it's called smishing.

One of the notable changes for financial institutions is that fraud on the Zelle® Network and other real-time channels add costs for products the market has effectively priced at free to the consumer. Fraud on the card networks, on the other hand, while occurring in real time and leveraging rules at the switch, can be priced into the offering, which provides interchange income to institutions. This makes it difficult to envision the benefits of a costly layered security posture when historically, consumers have been covered only for account take over – where the bad actor infiltrates the online or mobile banking system and moves money out without appropriate authorization.

To protect the institution, and its account holders, financial institutions, whether they like it or not, are going to be forced to rethink their security posture and the economics of instant payment services. They have to consider the possible loss of merchant interchange revenues to new competitive lower cost payment acceptance options to at least keep these merchants and have them use the cost savings to provide new loyalty incentives for their customers to come back and use loyalty for their next purchase.

To appropriately mitigate for different kinds of attack, institutions should begin to review their capabilities, and those of their vendors in the following areas:

- Negative Lists – Record bad actor lists that cross all payment methods and include account numbers, phone numbers, email addresses, IP addresses and other data elements that may be associated with them.

- Limits – Applicable limits should not only be at the transaction level, but also for velocity over time. One way to identify breakout fraud is to see the changes in payments in (or out) over time.

- Training - Have the right training in place in your call center to capture fraud/scam information at the initial point of contact.

- Transaction decisioning – Be prepared to review outbound transactions from the start. Then be prepared to add inbound transaction review to help identify whether you are banking bad actors.

- Transaction reporting to the networks – Gather the right data to report to the networks using the Federal Reserve Fraud Classifier model as your guide. Have a conversation with your processor or fraud vendor on how they can assist you with the workflow to gather and process this data quickly enough to meet network deadlines, and update it as required.

- Out-of-band confirmation – While on their own, One Time Passwords (OTP) are not without challenges, multi-factor authentication is an important part of a layered security posture. Consider ways to make this more secure, with appropriate messaging at the delivery point. Challenge your vendors to help you better serve your customers by keeping their codes safe.

- Harden account recovery – To reduce account takeover (ATO), institutions should review their password recovery procedures, both in the technical implementation in the digital platform, and in the call center. Call centers are often targets of social engineering and all employees should be trained in red flags.

- Step Up authentication - The best way to keep bad actors from coming through the front door is to add appropriate friction at account access. Institutions need upfront fraud review that looks at IP addresses, device and browser information as well as account holder behavior patterns, to reduce the chance of bad actors gaining access.

- KYC Validation - Validation processes should include verification of phone numbers and email addresses. Services are available that can identify whether an account holder is associated with a mobile plan, for example.

- Track non-reg E items - If you aren't doing so today, this is going to be a requirement soon enough. Best to get out ahead of it.

  ◉ Establish formal written policies on reimbursement for non-reg E items. Don't rely on per-item judgments.

- Fed Fraud Classifier - Implement fraud reporting mapped to the Federal Reserve Fraud Classifier Model. Identify the gaps and take them to your vendors. Investigate fraud mitigation solutions with machine learning capabilities, and consortium models.

- Financial institutions and payment services can and should also leverage AI to help combat fraud. It's an AI war.

Conclusion

With so much change happening so quickly, effectively navigating risk with real-time payments is going to require education and tenacity to keep up with rules and regulations. Institutions will have to mindfully identify risks and potential mitigations across:

- Compliance risk

- Cross-channel risk

- Fraud risk

- Liquidity risk

- Operational risk

- Strategic risk

- Systemic risk

- Reputational risk

- Technology risk

It will be important to keep in mind that the fraud/scam risks aren't new, but evolving, and the new rails don't mean starting from scratch, but identifying those differences that will impact policies, procedures, timing, and technology. Having a robust risk mitigation program in place, will enable your institution to better respond to ongoing changes, while keeping your larger payment strategy in mind.

Chapter 8 - Instant Payment rocket boosters: AI, QR codes, Open Banking APIs, Embedded Payments, and Directories

By Steve Wasserman

Major factors in the successful adoption and use of instant payments involve the use of these technologies and functional capabilities: In this chapter we take a brief look at the role that each of these play in instant payments.

Artificial Intelligence (AI)

AI has and will continue to affect our lives in many ways. We are only seeing the tip of the iceberg of how it plays a role in instant payments. To better understand the ways in which AI plays a role in instant payments, it will help to get an orientation of some of the types of AI and terminology related to it.

- Generative AI: This type of AI involves the creation of new content, such as images, text, or even music, imitating and creating original material. It doesn't just recognize patterns or make predictions based on existing data; it generates new data.

- Predictive AI: This form of AI uses historical data and machine learning to make predictions or forecasts about future events. It analyzes patterns in the data to anticipate potential outcomes.

- Interpretive AI: Interpretive AI aims to comprehend and interpret data in a manner that simulates human understanding. It involves reasoning and understanding the context to make sense of information.

- Extractive AI: This type of AI involves extracting specific information or data from a given source. It focuses on pulling out particular details or data points from documents or other sources without altering the original content.

- Large Language Models : LLM models, such as those used by Chat GPT and other types of AI described above, are AI systems capable of understanding and generating humanlike text. They can comprehend context and generate relevant responses based on the input they receive.

- Natural Language Processing: NLP is a branch of AI focused on enabling computers to understand, interpret, and generate human language. It involves language understanding, language generation, and other language-related tasks.

- Machine Learning: ML is an AI technique that allows systems to learn and improve from experience without being explicitly programmed. It enables algorithms to improve their performance over time.

- Big Data Analytics: It involves the analysis of large and complex data sets to identify patterns, correlations, and other insights. It helps in understanding trends and making informed decisions.

- Behavioral Analytics: This involves analyzing patterns of human behavior, especially in digital systems, to predict actions or detect anomalies based on deviations from normal behavior.

- Biometrics: Biometrics involves using physical characteristics (such as fingerprints, facial recognition, or iris scans) for identification or authentication purposes.

- Robotic Process Automation: RPA uses software robots or "bots" to automate repetitive tasks, enabling the automation of processes that mimic human interactions with digital systems.

AI's use in instant payments offers a range of benefits, including:

- Enhanced user experiences such as assisted data entry, mistake detection and correction, and predictive suggestions

- Fraud detection, prevention, and mitigation

- Automated payment processing such as
  - Payment initiation
  - Payment reconciliation
  - Automatic directory lookups, additions, and updates

- Cost savings by reducing errors and ensuring compliance

Embracing AI technologies in payment processing workflows is crucial for maintaining competitive advantages in the realm of instant payments.

Quick Response Codes (QR Codes)

QR codes are two-dimensional barcodes capable of storing more information than a traditional barcodes and can be read faster by scanning devices[1]

In the realm of instant payments, QR codes serve several purposes:

- QR codes for point-of-sale payments: Merchants utilize QR code scanners to facilitate payments from a customer's mobile app or vice versa.[1]

- Mobile-initiated transfers: Mobile wallets and money movement apps incorporate QR codes for person-to-person fund transfers.[1]

- QR codes on bills and invoices: Billers and suppliers can employ QR codes on paper bills, PDFs, or on their website. Scanning apps can use these codes to initiate payments or link to an e-commerce payment platform.[1]

- QR codes on restaurant receipts: Alongside paperless menus, restaurants use QR codes on the 'check' for contactless payment and checkout-at-the-table experiences.[1]

- Loyalty programs at checkout: Grocery stores and similar businesses allow customers logged into an app to present a QR code that integrates loyalty/rewards with a payment method for POS, unattended checkout, or online payments.[1]

QR codes can be integrated with directories containing verified and updated payment information. By scanning a secure directory lookup ID from the QR code, an API can retrieve the latest information, ensuring accuracy and enhanced security, as opposed to directly obtaining payment information from the QR code itself.

If payment account information is to be scanned from a QR code, it should, at a minimum, be tokenized or encrypted. The instant payment network, service, or scanning app must be capable of translating tokenized or encrypted information into the appropriate account details.

QR codes have significantly influenced the widespread adoption of instant payments in various countries, such as those in Asia, India, Brazil, and others. They have transformed retail outlets and marketplaces, sometimes becoming the preferred method of payment initiation. In certain street markets, loading funds into local payment services that only accept QR code payments has become common, as these services often avoid costly card and cash transactions which can make them vulnerable to potential theft when closing for the day.

Undoubtedly, the use of QR codes in instant payments serves as a significant adoption catalyst based on its easy to use, lower cost, and safer transactions.

Instant Payment APIs, Open Banking, and Embedded Payments

An API, or Application Programming Interface, is a set of rules, protocols, and tools that allows different software applications to communicate and interact with each other. APIs define the methods and data formats that applications can use to request and exchange information, enabling seamless integration between diverse systems or services.

APIs enable real-time functions as a natural fit for the shift from batch file-based payment networks, such as ACH to real-time instant payments.

Instant payments networks clear and settle individual transactions through a process called Real Time Gross Settlement (RTGS). These networks enable the exchange and processing of instant payment ISO 20022 messages with their network participants. The real-time access and response capabilities of APIs are just a natural fit for instant payment initiation and other ISO 20022 functions.

Open Banking APIs go beyond the functions of what the ISO 20022 messages enable. They offer a wide range of functionalities that can transform how financial services are accessed and utilized. Some of the key functions and capabilities enabled by Open Banking APIs include:

- Account Information Services: AIS allows users to aggregate their account information from multiple banks or financial institutions in one place. This enables them to view account balances, transaction history, and other account details in a consolidated manner as well as be able to electronically share that information with other parties, such as when opening a new account.

- Payment Initiation Services (PIS): PIS allows users to initiate payments directly from their bank accounts without the need to use traditional payment methods. This can streamline transactions and enable direct payments between different parties. The use of ISO 20022 messaging is most applicable here.

- Identity Verification and Authentication: Open Banking APIs can be utilized to facilitate secure identity verification and authentication processes, enhancing the security of financial transactions and services.

- Credit Scoring and Loan Applications: By accessing an individual's financial data (with their consent), Open Banking APIs can enable lenders to make more informed credit decisions and offer personalized loan products based on the user's financial history.

- Personal Financial Management (PFM) Tools: Open Banking APIs allow for the development of applications that provide budgeting, expense tracking, and financial planning tools by leveraging a user's transactional data from multiple accounts.

- Third-Party Financial Services Integration: APIs enable third-party developers to build new financial products and services, integrating them into existing banking systems, providing customers with a more diverse range of options.

- Compliance and Regulatory Services: Open Banking APIs can help financial institutions stay compliant with regulations by providing standardized access to data and ensuring that data sharing adheres to regulatory requirements.

- Marketplace and Comparison Services: They enable the creation of platforms that allow users to compare financial products and services from different providers, fostering competition and giving users more options.

- Risk Assessment and Fraud Prevention: APIs can be used to assess risks and help in the prevention of fraud by analyzing patterns and anomalies in financial data.

These functionalities empower customers with more control over their financial data and enable banks, fintech companies, and other financial service providers to offer more innovative and personalized services, ultimately improving the overall banking experience.

Businesses can use these Instant payment and Open Banking APIs provided by their financial institutions and third-party services to embed instant payment functionality into their business systems and applications that they offer to their clients, customers, and partners.

A great example of embedded instant payments are those used by Uber and Lyft where you have pre-authorized access to charge your account or push your payment as you end the ride. The driver in turn can also have the share of the transaction and tip paid to them using instant payments APIs on the back end of each ride or when they check out of their time driving for one of these services. Amazon checkout experiences and other loyalty/customer id related check out experiences are also good examples of embedded payments.

Directories

Directories play a crucial role in instant payments by facilitating the use of aliases or other identification provided by the payee to lookup payment information to use to send them an instant payment. The directory can also be used to lookup the payer's account information to send them an instant request for payment or otherwise debit their account through an ACH or off of their card account.

As per the US Faster Payments Council Directory Models Work Group's infographic "Making the case for an Interoperable Directory to facilitate Faster Payments"[2] which is based on their whitepaper deliverables about the characteristics and economic benefits of an interoperable faster payments directory, it summarizes the fundamental attributes of as:

- Safety

- Interoperability

- Governance

The characteristics of an interoperable directory are:

- Accurate Routing

- Unique Alias

- Support Multiple Payment Routes

- Minimized Storage of Sensitive Data

- Data Mining Safeguards

- End User Controlled Profiles

- Payee Verification

- Request for Payment Support

The economic benefits derived from using directories with faster payments were estimated and summarized as:

- Accuracy in Routing: $188 million in savings of misrouted payments

- Diminished Fraud: $1.7 billion in email fraud losses and $200 million in losses to fake entities

- Consumer Confidence in Aliasing: $242 in revenue as per increased volume of UPI payments in India

- Compliance: $90 million in compliance savings, just for utility companies

- Automation: $91 billion in cost savings through automated decisioning

- Electronification: $13 billion in cost savings for moving B2B checks to electronic

- User Experience: reduced late fees and more satisfied users

Overall, directories play a pivotal role in instant payments by serving as a foundational infrastructure that supports the secure, fast, and reliable lookup of account information to use for instant payment transactions. This helps foster adoption of instant payments and realize the subsequent economic benefits.

Conclusion

When you put all this together on the front end, you may scan a QR code and pass a secure id to a directory using an API. Then using an API to verify the payment account and another API to initiate the transaction or send the payer a request for payment.

AI plays a part in this flow either embedded within one of the APIs or running in the background somewhere through the end-to-end flow. AI may check the payee identification as one which may have been associated with or suspected fraudulent activity. It can analyze the payer's and payee's transaction history to identify abnormalities or possible errors it can warn about through a status returned through the API, as part of the user interface, or on the back end as the transaction passes through the payment service, FI, and/or network.

AI can also automatically extract the biller or payee off of a bill/invoice and look up the directory id to start with. Then it can provide the default amount and date to be scheduled for payment the day or moment it is due. It can further predict if your cash flow forecast will have sufficient funds at that time the payment is scheduled for.

There are numerous other scenarios we could imagine how all of these things can provide value on their own as well as in conjunction with each other.

The bottom line on all of this is that these value-added functionalities serve as "rocket boosters" for the adoption of instant payments.

Conclusion

In today's financial landscape, the concept of instant payments has emerged as a pivotal and revolutionary system with far-reaching implications. Understanding the depth of what instant payments truly entails is crucial in comprehending its significance and the transformative impact it holds across social, economic, and philosophical realms.

The history of instant payments, both on a global scale and within the United States, is a story of volution and adaptation to meet the growing demands for faster, more efficient financial transactions. At the heart of this evolution lies the ISO 20022 standard, a cornerstone that defines the features, functionalities, and overarching benefits of this innovative payment system.

The deployment of instant payments unveils a multitude of use cases and their corresponding implementation considerations. From enhancing customer experiences to reshaping business strategies, the implementation of instant payment systems need to address risks, fraud, and compliance considerations, redefining the operational landscape for financial institutions, payment services, and their end consumer and business customers.

To foster adoption and enhance instant payments, various technological advancements like AI, QR code technology, Open Banking APIs, Embedded Payments, and directories play instrumental roles.

Waiting for customers to demand instant payments might result in missed opportunities and potential migration of customers to other institutions that offer such services.

For institutions and payment services yet to embark on the instant payments journey, establishing a roadmap and timeline is imperative. Evaluating what works, what doesn't, and planning future phases is essential for those already on this transformative path.

Considering use cases, strategies, risks, and compliance within a roadmap becomes a critical component. Mastering and leveraging the capabilities and rich data of ISO 20022 messages is an essential aspect of the value proposition of your use cases.

Collaboration and information access through industry associations, Open Banking, directories, and fraud information sharing have a network effect that you can never achieve trying to go it alone.

The ultimate goal of instant payments is ubiquity, where transactions can be sent and received between any two accounts in the US and eventually, globally. Achieving instant payment ubiquity would mean not only speed, but also interoperability across various payment networks.

The future of instant payments holds promise. It envisions a world where nearly everything is instant or at least faster, akin to the ease and speed we've grown accustomed to with online purchases and deliveries. This includes the widespread use of QR codes, directories that eliminate the need for account numbers, and financial inclusion for the unbanked and underserved.

In the forthcoming years, instant payments will continue to evolve, incorporating newer forms such as digital currencies and digital accounts and expanding into various sectors. This includes the migration of credit and debit cards into virtual, secure, and instantly settled transactions. Additionally, AI will be instrumental in enhancing user experiences and automating various aspects of the payment ecosystem.

Reflecting back in five or ten years, this pivotal period of embracing instant payments will serve as a transformative chapter in the financial landscape. This journey is about adaptation, innovation, and reaping the benefits of a faster, more efficient financial ecosystem. See you on the other side!

Contributors

We want to thank and recognize all of the contributors to this book that included the following professionals in the payments industry.

Elspeth Bloodgood, AAP, NCP, Chapter Contributor

Elspeth is a senior product manager for Jack Henry's JHA PayCenter and a subject matter expert in the Zelle, FedNow, and RTP networks. She has been in the payment industry for the better part of 20 years, starting with a stint on the x9B committee that created the IRD standard for Check 21. She has been a frequent work group member for NACHA,

the Faster Payments Council and its predecessors, and other cross industry groups. She has experience in online BillPay for the iPay Division of Jack Henry. Prior to that, she was in product management for biller direct and ACH products at multiple companies.

Attila Csutak, Chapter Contributor

Attila Csutak's career spans 24 years in Treasury Management practice with Large Financial Institutions (top 5 domestic and international) and a large Fintech company. Attila held product development, product management and product strategy roles. His knowledge of all payment types and systems provided the background required for kick starting the RTP efforts at a Fintech company and was deeply involved in two major acquisitions the Fintech company executed to address the growing needs of Faster Payments needs of Financial Institutions in the U.S.A. Integral contributor to the development of the Real Time Payments idea submitted to the FIN Tank (Fiserv Innovation Network) to deliver B2B Real-Time Payments to nearly 6,000 Financial Institutions via The Clearing House (TCH).

Proven track record in strategic planning, building multigenerational product plans, and influencing customer portfolios to grow the product. Experienced in market planning, product strategy and the establishment of teams to optimize customer on-boarding and marketing strategies. Involved in numerous industry collaborations, consultations, and speaking events.

Currently Payables Team Lead (Legacy and Faster Payments) with City National Bank, serving as evangelist for Emerging Payments

Peter Davey, Chapter Contributor

Peter is an accomplished executive, industry thought leader, influencer and strategist across the disciplines of payments, banking and financial services.

He is currently a Venture Partner at Alloy Labs focused on building out platforms that help community banks and credit unions succeed in payments and banking.

Previous to Alloy, Peter was Head of Product Innovation and Labs at The Clearing House. He is credited as one of the original architects involved in designing and launching RTP® from The Clearing House, the first real-time payment system in the US.

Prior to that, he spent ten years as Head of Payment Strategy, Innovation & Industry at Capital One, where he developed and implemented a federated payments governance model for the company, and established the Enterprise Payments Advisory Council, a senior executive council to facilitate cross-functional governance across business lines.

He has a proven track record of building teams & organizations, establishing payments governance, building consensus around payment strategies and providing subject matter expertise and thought leadership in many industry forums.

Travis Dulaney, Chapter Contributor

Travis is a dynamic leader known for guiding banks and fintechs through complex challenges in payments, banking, risk management, and compliance. His impressive journey includes founding BalancedTrust, a Third & Fourth Party Risk Management venture for banks, and co-founding The Ethica Group, a Venture Builder supporting entrepreneurs in realizing their visions.

As the visionary Founder & CEO of PayFi, Travis pioneered a groundbreaking payment processing venture that connected Community Banks to real-time payment schemes, culminating in a successful exit in September 2020.

Travis's career boasts pivotal roles at FIS, including overseeing the FDIC contract during the 2008-2012 economic crisis, liquidating over 2500 insolvent banks. His expertise spans software development, Six Sigma process engineering, and a deep understanding of the payment ecosystem, including Identity Management, Acquiring, Merchant Processing, and Card Networks.

Travis's superpower lies in translating knowledge into innovative processes. His strategic acumen in Banking Operations, Risk Management, and Compliance makes him an unparalleled authority. His influence extends to ACH, Real-Time Payments, Credit Cards, and emerging technologies like Digital Assets, along with Banking As A Service.

Reed Luhtanen, Chapter Contributor

Reed is Executive Director of the U.S. Faster Payments Council (FPC). The FPC is an industry-led membership organization whose mission is to facilitate a world-class payment system where

Americans can safely and securely pay anyone, anywhere, at any time and with near-immediate funds availability. Reed is responsible for managing the daily operations of the organization and working with the FPC board and membership to execute on the FPC's strategic plan while ensuring inclusive and transparent dialogue with all FPC stakeholders.

Reed has extensive experience in the payments industry and has served on myriad industry bodies, most recently an officer on the FPC Board of Directors; the Federal Reserve-sponsored Governance Framework Formation Team; and the Corporate Advisory Group for The Clearing House's Real-Time Payments System. Prior to the FPC, he spent 15 years at Walmart as senior director of global treasury.

In 2015, Reed was recognized as one of PayBefore's "Top Ten Payments Lawyers" and The Electronic Transactions Association honored him as a member of its "Forty Under 40" class in 2019.

Kevin Olsen, AAP, NCP, APRP, CHPC, "Payments Professor", Chapter Contributor

Kevin is a renowned expert and educator in the field of payments. With over two decades of experience, he is known as the "Payments Professor" for his ability to simplify complex concepts and make them accessible to anyone.

As a speaker, Kevin has a unique talent for engaging his audience with his passion for the subject matter. He has been invited to speak at industry conferences, academic institutions, and corporate events around the world.

Kevin is highly respected in the payments industry for his insights on

emerging technologies, regulatory issues, and consumer behavior. He has consulted for financial institutions and technology companies, providing strategic guidance on product development and market positioning. He is the SVP, of Innovation and Strategy for Pidgin, a product leading the way for instant payments in the U.S.

Kevin holds a master's in industrial organizational psychology and is a frequent commentator in the media. He has an NCP Certification and is an NCP Certified Trainer as well as Accredited Payments Risk Professional (APRP), Microsoft Certified System Engineer (MCSE), Microsoft Certified System Administrator (MCSA), and Microsoft Certified Professional (MCP) certifications.

245

Sean Rodriguez, Foreword Contributor

Sean Rodriguez retired from the Federal Reserve System in January 2019. Sean left the Federal Reserve as Executive Vice President and Faster Payments Strategy Leader. As Faster Payments Strategy Leader, Rodriguez led activities to identify effective approaches for implementing a safe, ubiquitous, faster payments capabilities in the United States. In addition, Rodriguez chaired the Federal Reserve's Faster Payments Task Force, comprised of more than 300 payment system stakeholders interested in improving the speed of authorization, clearing, settlement and notification of various types of personal and business payments.

Rodriguez retired after more than 35 years with the Federal Reserve in operations, product development, sales, marketing and bank administration at the Denver, Los Angeles and Chicago locations. He helped establish the Federal Reserve's Customer Relations and Support Office in 2001 including its national account program and served on the Federal Reserve's leadership team for implementing the Check 21 initiative. More recently, Rodriguez was instrumental in the design and launch of the Federal Reserve's Payments Industry Relations Program. Rodriguez holds a B.A. from the University of Colorado, a Graduate School of Banking Certificate from the University of Wisconsin and is an Association for Financial Professionals - Certified Cash Manager.

246

Connie Theien, Chapter Contributor

Connie Theien is a payments expert with more than 20 years experience leading payment system improvement programs as a Federal Reserve executive. She retired in December 2023 as senior vice president and head of industry relations, a function she worked to establish in 2013 with an aim to engage a broad swath of industry stakeholders in advancing U.S. payment system improvements.

Theien's career spanned payments industry transformation programs encompassing ACH adoption, Check 21 implementation, the introduction of cross-border and same-day ACH payments, and instant payments catalyzation and implementation.

Theien led collaboration and education programs focused on faster payments, fraud prevention, and efficiency improvements for B2B and cross-border payments. She designed and facilitated the Federal Reserve's Faster Payments Task Force and follow-on efforts to drive adoption of faster payments in the United States, in addition to playing a central role in implementation of the FedNow instant payments service.

Prior to her tenure at the Fed, Theien led marketing and public relations efforts for nonprofit, retail and academic organizations. Theien holds a B.A. in communications and an M.B.A. from the University of Minnesota.

Steven Wasserman, Chapter Contributor and Overall Book Content Editor

Steve co-founded a successful Fintech company, IPP, which enabled financially inclusive walk-in bill payment plus prepaid & debit card solutions. A series of acquisitions and name changes ended with a PayPal acquisition for over $300m. Steve then founded Vments,

commercializing tokenized deposits and Enterprise Digital Banking (EDB) Platform.

Now at Photon

Commerce, Steve helps clients and partners leverage AI software to automate financial documents. Use cases include enhanced and automated front end payment initiation,

backend automated

straight through processing, and everything in-between helping to deliver instant payment user experiences, and enabling directories for document exchanges.

For decades Steve consulted ERP clients integrating payables and receivables payments and other streamlined automation between banks and trading partners. Steve helps banks and fintechs strategize to leverage faster payments and the latest technologies as part of their roadmaps to "Banking Into The Future".

Steve provides leadership in committees at the US Faster Payments Council (FPC), the Fed Improvements facilitated Business Payments Coalition, X9 ISO 20022 Market Practices Forum, the FedNow Community, and the FPC Board Advisory Group.

Michael Young, Publisher & Editor

Michael Young is an entrepreneur in Silicon Valley and the founder & CEO of Photon Commerce. After leaving a Stanford PhD, Michael was an engineer at IBM prior to starting 3 venture-backed startups. He led advanced R&D programs funded by the Department of Defense, created hundreds of jobs, raised over $50M across his venture-backed startups, and took some to acquisitions. Photon Commerce's payments AI and document understanding technology enables 1-click bill pay.

Photon Commerce provides white-labeled multi-layered AI-as-a- Service to fintech leaders, enabling them to bring new instant product experiences to market. Photon Commerce is like Plaid for B2B documents, enabling platforms to launch Generative Pretrained AI/ OCR to the market in as fast as 1 day.

Photon Commerce is supported by the AngelList Quant Fund and Village Global, a venture firm funded by Jeff Bezos and Bill Gates.

References

Introduction

1. The US Faster Payments Council Education and Awareness Work Group's Glossary Sub-group, https://fasterpaymentscouncil.org/

2. https://www.frbservices.org/financial-services/fednow/about.html

3. https://bootcamp.cvn.columbia.edu/blog/what-is-fintech/

4. Epcor.org, sourced from Fed FP Task Force

5. ISO 20022.org, https://www.iso20022.org/

6. Nacha, Glossary of International ACH Terms, https://www.nacha.org/content/international-payments-glossary

7. Fed Payments Improvements, https://fedpaymentsimprovement.org/

8. https://fasterpaymentscouncil.org/Overview

9. https://www.zellepay.com/faq/what-zelle

Chapter 1 - Societal, Economic, and Philosophical Implications of Instant Payments

1. https://www.federalreserve.gov/newsevents/speech/brainard20190805a.htm

2. You can hear Professor Swartz discussing her book and her thoughts on payments as social media on the February 9, 2023 episode of "Off the Rails from the U.S. Faster Payments Council" podcast

3. New Money: How Payment Became Social Media, Lana Swartz, pg 4

4. A History of Communications: Media and Society from the Evolution of Speech to the Internet, Marshall T. Poe, pg X

5. The Greek poet, not Bart's dad

6. Id, at pg 7

7. Id, at pg 12

8. Id, at 83

9. Id, at 107

10. Id, at 113

11. Id, at 120

12. Later in this chapter we will revisit this brief timeline to see where various payment methods fall

13. No doubt some readers got here and thought "hey, what about radio and TV?" For sure, both are important, and both are covered in Poe's book. For these purposes, I decided to limit the analysis to written word media, which in my view are most analogous to payments as a form of media.

14. Poe at 215

15. Id at 217

16. "Death Wish" by Jason Isbell and the 400 Unit

17. https://www.investopedia.com/articles/07/roots_of_money. asp#:~:text=In%20600%20BCE%2C%20Lydia's%20 King,pictures%20that%20acted%20as%20denominations.

18. https://fin.plaid.com/articles/checking-out-a-brief-history-of-checks/#:~:text=The%20first%20checks%20started%20cropping,hand%2C%20sort%20of%20like%20IOUs.

19. Payments Systems in the United States, Carol Coye BHenson, Scott Loftesness, Russ Jones, at 29,

20. https://www.fdic.gov/analysis/household-survey/index.html

21. https://www.atlantafed.org/-/media/documents/banking/consumer-payments/research-data-reports/2020/02/13/us-consumers-use-of-personal-checks-evidence-from-a-diary-survey/rdr2001.pdf

22. It's certainly worth noting that card payments also emerged in the middle of the 20th century and have certainly had a transformative impact on the way Americans pay for things. That said, because of the obvious parallels between ACH and instant payments, I'm choosing to focus on ACH here in this chapter. No doubt, entire books could be written on the card networks.

23. Payments Systems in the United States, Carol Coye BHenson, Scott Loftesness, Russ Jones, at 48,

24. https://www.nacha.org/content/ach-network-volume-and-value-statistics

25. Remarks made by Mr. Coaxum during the "Doing Well by Doing Good" Panel Presentation at the U.S. Faster Payments Council's 2022 Spring Member Meeting,

26. Faster Payments and Financial Inclusion, U.S. Faster Payments Council, https://fasterpaymentscouncil.org/blog/9874/Faster-Payments-and-Financial-Inclusion-White-Paper

27. according to BankRate: https://www.cbs19news.com/story/34248451/6-in-10-americans-dont-have-500-in-savings

28. https://www.fdic.gov/analysis/household-survey/index.html

Chapter 2 - History of Instant Payments

1. Investopedia,

2. Martin Mayer, "The Humbling of BankAmerica" The New York Times, May 3,1987, p.27,

3. History of NACHA and the ACH Network, April 20, 2019, NACHA, www.NACHA.org/content/history-NACHA-and-ACH-network

4. Payment System Improvement -Public Consultation Paper,September 10, 2013,Federal Reserve , https:// fedpaymentsimprovement.org/wp-content/uploads/2013/09/ Payment_System_Improvement-Public_Consultation_Paper. pdf

5. US Faster Payments Council Education & Awareness Glossary of Terms Work Group collaboration,

6. Federal Reserve announces posting rules for new same-day automated clearing house service, June 21, 2010, https://www. federalreserve.gov/newsevents/pressreleases/other20100621a. htm

7. Federal Reserve Board approves enhancements to Reserve Banks' same-day ACH service, September 23, 2015, https://www.federalreserve.gov/newsevents/pressreleases/ other20150923a.htm

8. Speech given by Sandra Pianalto, President of the Federal Reserve Bank of Cleveland, October 22, 2012, https:// fraser.stlouisfed.org/files/docs/historical/frbclev/presidents/ pianalto_20121022.pdf

9. Federal Reserve Banks laud industry response to consultation paper and highlight next steps, February 04, 2014, https:// www.federalreserve.gov/newsevents/pressreleases/ other20140204a.htm

10. Strategies for Improving the U.S. Payment System, January 26, 2015, https://fedpaymentsimprovement.org/wp-content/uploads/strategies-improving-us-payment-system.pdf

11. Strategies for Improving the U.S. Payment System: Federal Reserve Next Steps in the Payments Improvement Journey, September 6, 2017, https://fedpaymentsimprovement.org/wp-content/uploads/next-step-payments-journey.pdf

12. Get to Know the Faster Payments Task Force, https://fedpaymentsimprovement.org/news/blog/get-to-know-the-faster-payments-task-force/

13. Federal Reserve announces steering committees of new Payments Task Forces, July 21, 2015, https://www.federalreserve.gov/newsevents/pressreleases/other20150721a.htm

14. Learn About the Phases of the Faster Payments Task Force Work, https://fedpaymentsimprovement.org/news/blog/learn-about-the-phases-of-the-faster-payments-task-force-work/

15. Federal Reserve's Secure Payments Task Force identifies key priorities, seeks comments, https://www.federalreserve.gov/newsevents/pressreleases/other20161025b.htm

16. Faster PaymentsEffectiveness Criteria, January 26, 2016, https://fedpaymentsimprovement.org/wp-content/uploads/fptf-payment-criteria.pdf

17. Faster Payments Journey, https://fedpaymentsimprovement.org/strategic-initiatives/faster-payments/journey/

18. The U.S Path To Faster Payments, Final Report Part Two: A Call to Action, July 2017, https://fedpaymentsimprovement.org/wp-content/uploads/faster-payments-task-force-final-report-part-two-2.pdf

19. The U.S Path To Faster Payments, Final Report Part One: The Faster Payments task Force Approach, January 2017, https://www.federalreserve.gov/newsevents/press/other/us-path-to-faster-payments-pt1-201701.pdf

20. U.S. Faster Payments Council Website https://fasterpaymentscouncil.org/Overview

21. The Clearing House makes history with new US payments system, November 14, 2017, https://www.fintechfutures.com/2017/11/the-clearing-house-makes-history-with-new-us-payments-system/

22. RTP® Network Surpasses Half a Billion Instant Payments, July 24, 2023, https://www.theclearinghouse.org/payment-systems/Articles/2023/07/07-24-2023_RTP_Network_Surpasses_Half_Billion_Instant_Payments

23. RTP® Network Surpasses 1-Million Payments on a Single Day, September 06, 2023, https://www.theclearinghouse.org/payment-systems/Articles/2023/09/RTP_Network_Surpasses_1_Million_Payments_Single_Day_09-06-2023

24. Potential Federal Reserve Actions To Support Interbank Settlement of Faster Payments, Request for Comments, November 15, 2018, https://www.federalregister.gov/documents/2018/11/15/2018-24667/potential-federal-reserve-actions-to-support-interbank-settlement-of-faster-payments-request-for

25. A New 24x7x365 Instant Payment Service https://fedpaymentsimprovement.org/strategic-initiatives/faster-payments/overview/

26. Federal Reserve Actions To Support Interbank Settlement of Faster Payments, August 9, 2019, https://www.federalregister.gov/documents/2019/08/09/2019-17027/federal-reserve-actions-to-support-interbank-settlement-of-faster-payments

27. Service Details on Federal Reserve Actions To Support Interbank Settlement of Instant Payments, August 11, 2020, https://www.federalregister.gov/documents/2020/08/11/2020-17539/service-details-on-federal-reserve-actions-to-support-interbank-settlement-of-instant-payments

28. Announcing the FedNowSM Pilot Program participants, January 25, 2021, https://www.frbservices.org/financial-services/fednow/community/news/012521-announcing-pilot-program-participants.html

29. Number of real-time payments (RTP) transactions in 44 countries and territories worldwide in 2022, with a forecast for 2027, Statista https://www.statista.com/statistics/1422323/real-time-payments-forecast-by-country/

Chapter 3 - Understanding ISO 20022: The Universal Language of Financial Messaging

1. About ISO 20022, https://www.iso20022.org/about-iso-20022

2. The Banker, October 11, 2022, https://www.thebanker.com/ISO-20022-A-love-story-1665474464

3. ISO 20022 Migration for European Banks, Accenture, https://www.accenture.com/content/dam/accenture/final/industry/banking/document/Accenture-Banking-Europes-Journey-ISO20022-Migration.pdf

4. Switzerland looks to the future with new ISO 20022 payments architecture, Finastra, April 18, 2016, https://www.finextra.com/newsarticle/28751/switzerland-looks-to-the-future-with-new-iso-20022-payments-architecture

5. SwissSTPC Comments, Accenture, October 7 2022, https://www.six-group.com/dam/download/sites/swiss-sptc/iso-20022-accenture-study-swiss-sptc-comments.pdf

6. Navigating the Evolution: Nordic Banking in Transition from SWIFT FIN to ISO 20022, Tietoevcy, December 12, 2023, https://www.tietoevry.com/en/blog/2023/12/nordic-banking-in-transition-from-swift-fin-to-iso-20022/#:~:text=In%20a%20significant%20industry%20shift,industry%2Dwide%20transition%20to%20ISO20022.

7. High-value payment system – Lynx, Payments Canada, https://www.payments.ca/payment-resources/iso-20022/high-value-payment-system-lynx

8. ISO 20022 Migration for the Australian Payments System – Issues Paper, Reserve Bank of Australia and the Australian Payments Council, 2019, https://www.rba.gov.au/publications/consultations/201904-iso-20022-migration-for-the-australian-payments-system/pdf/issues-paper.pdf

9. ISO-20022 Adoption Study for The Monetary Authority of Singapore, Monetary Authority of Singapore, https://www.mas.gov.sg/-/media/MAS/Singapore-Financial-Centre/Why-Singapore/MEPS/References/MAS-SWIFT-ISO-20022-Adoption-Study.pdf

10. World Bank FAST Payments Toolkit Case Study, Hong Kong SAR, China, The World Bank, 2018, https://fastpayments.worldbank.org/sites/default/files/2021-09/World_Bank_FPS_Hong_Kong_SAR_China_FPS_Case_Study.pdf

11. Version Upgrade of the ISO 20022 Messaging Standard for the BOJ-NET, Bank Of Japan, January 21, 2023, https://www.boj.or.jp/en/paym/bojnet/net_forum/nfo230131a.pdf

12. ISO 20022 Post go-live media release, South African Reserve Bank, November 3, 2022, https://www.resbank.co.za/content/dam/sarb/publications/media-releases/2022/iso-20022/ISO%2020022%20Post%20go-live%20press%20release.pdf

13. Brasilan Payments Oversight Report, Banco Central Do Brazil, 2015, https://www.bcb.gov.br/content/financialstability/paymentssystem_docs/Payments-System-Oversight-Report/Brazilian-Payments-System-Oversight-Report-2015.pdf

14. Why ISO 20022, Swift, https://www.swift.com/standards/iso-20022/supercharge-your-payments-business/chapter-2#:~:text=ISO%2020022%20provides%20a%20common,processes%2C%20and%20ultimately%20cash%20reporting.

15. About ISO 20022, ISO 20022.org, https://www.iso20022.org/about-iso-20022#:~:text=financial%20standards%20initiatives-,ISO%2020022%20is%20a%20multi%20part%20International%20Standard%20prepared%20by,transactions%20and%20associated%20message%20flows

16. MT to ISO 20022: The challenge of modernizing your payment messages, Trade Header, https://www.tradeheader.com/blog/mt-to-iso-20022-the-challenge-of-modernizing-your-payment-messages#:~:text=ISO%2020022%20is%20richer%2C%20better,the%20automation%20of%20its%20processing.&text=%E2%80%9CMT%20messages%20are%20very%20simple,industry%20for%20a%20long%20time.

17. ISO 20022 vs ISO 8583: What is the Difference?, IR, https://www.ir.com/guides/iso-20022-vs-iso-8583

18. ISO 20022 for Payments for Financial Institutions, SWIFT, https://www.swift.com/standards/iso-20022/iso-20022-payments-financial-institutions

Repeat

19. Fedwire® Funds Service ISO® 20022 Implementation Center, FRBservices.org, https://www.frbservices.org/resources/financial-services/wires/iso-20022-implementation-center

20. TCH Reschedules CHIPS ISO 20022 Implementation to April 2024, The Clearing House, February 9, 2023, https://www.theclearinghouse.org/payment-systems/Articles/2023/02/02-09-2023_TCH_Reschedules_CHIPS_ISO_20022_Implementation_April_2024

Chapter 4 - Implementations: Strategic Approach & Go To Market Plan

1. Travis Dulaney Experience, BalancedTrust, Inc. 2023, https://www.balancedtrust.co/

Chapter 5 - Introduction to instant payment use cases and their benefits

1. Bank Innovation. (2021). The Future of Payments: Instant Payments. Retrieved from https://bankinnovation.net/2021/06/the-future-of-payments-instant-payments/

2. Bank Innovation. (2021). The Future of Payments: Request for Payment. Retrieved from https://bankinnovation.net/2021/07/the-future-of-payments-request-for-payment/

3. Capgemini. (2021). World Payments Report 2021. Retrieved from https://www.capgemini.com/resources/world-payments-report-2021/

4. Consumer use cases for real-time payments. (2021). PYMNTS. Retrieved from https://www.pymnts.com/news/faster-payments/2021/consumer-use-cases-real-time-payments/

5. Danish National Bank. (2020). Payment Habits in Denmark. Retrieved from https://www.nationalbanken.dk/en/ publications/Pages/2020/05/Payment-habits-in-Denmark.aspx

6. Danish National Police. (2020). Statistics on Bank Robberies. Retrieved from https://www.politi.dk/en/statistics

7. Deloitte. (n.d.). Instant Payments and the Gig Economy. Retrieved from https://www2.deloitte.com/content/dam/ Deloitte/xe/Documents/financial-services/me_instant- payments-and-the-gig-economy.pdf

8. Deloitte. (n.d.). Instant Payments: Are Corporate Treasurers Ready? Retrieved from https://www2.deloitte.com/content/ dam/Deloitte/lu/Documents/financial-services/lu-instant- payments-are-corporate-treasurers-ready.pdf

9. European Central Bank. (2018). Instant Payments: The Future of Payments? Retrieved from https://www.ecb.europa.eu/ explainers/tell-me-more/html/instant_payments.en.html

10. European Central Bank. (2019). Instant Payment Systems: The New Normal. Retrieved from https://www.ecb.europa.eu/ press/key/date/2019/html/ecb.sp191128~28e52cbbf4.en.html

11. European Central Bank. (2020). Cash in the time of COVID-19. Retrieved from https://www.ecb.europa. eu/pub/economic-bulletin/focus/2020/html/ecb. ebbox202008_01~d8f7c5bae0.en.html

12. EY. (n.d.). Real-time payments adoption in Asia Pacific. Retrieved from https://www.ey.com/en_sg/banking-capital- markets/real-time-payments-adoption-in-asia-pacific

13. Federal Reserve. (2019). The State of Cash: Preliminary Findings from the 2019 Diary of Consumer Payment Choice. Retrieved from https://www.frbsf.org/cash/publications/fed- notes/2019/november/the-state-of-cash-preliminary-findings- 2019-diary-consumer-payment-choice/

14. Finder UK. (n.d.). A guide to instant payments in the UK. Retrieved from https://www.finder.com/uk/instant-payments

15. FIS Global. (n.d.). 5 Use Cases for Real-Time Payments. Retrieved from https://www.fisglobal.com/-/media/fisglobal/Files/PDF/whitepaper/real-time-payments-use-cases.pdf

16. FIS Global. (n.d.). How Request for Payment is Changing the Way We Pay. Retrieved from https://www.pymnts.com/news/faster-payments/2021/how-request-for-payment-is-changing-the-way-we-pay/

17. FIS Global. (n.d.). Instant Payments: Everything You Need to Know. Retrieved from https://www.fisglobal.com/-/media/fisglobal/Files/PDF/whitepaper/instant-payments-everything-you-need-to-know.pdf

18. FIS Global. (n.d.). Request for Payment: A New Era of Payment Collection. Retrieved from https://www.fisglobal.com/-/media/fisglobal/Files/PDF/whitepaper/request-for-payment.pdf

19. Finder UK. (n.d.). A guide to instant payments in the UK. Retrieved from https://www.finder.com/uk/instant-payments

20. How Request for Payment is Changing the Way We Pay. (2021). PYMNTS. Retrieved from https://www.pymnts.com/news/faster-payments/2021/how-request-for-payment-is-changing-the-way-we-pay/

21. Instant Payments in Asia Pacific: Ready for Prime Time. (n.d.). McKinsey & Company. Retrieved from https://www.mckinsey.com/business-functions/risk/our-insights/instant-payments-in-asia-pacific-ready-for-prime-time

22. Instant Payments in the UK. (n.d.). Payments UK. Retrieved from https://www.paymentsuk.org.uk/our-work/instant-payments-uk

23. McKinsey & Company. (2021). Global Payments Report 2021. Retrieved from https://www.mckinsey.com/industries/financial-services/our-insights/global-payments-report-2021/

24. McKinsey & Company. (2021). Instant Payments in Asia Pacific: Ready for Prime Time. Retrieved from https://www.mckinsey.com/business-functions/risk/our-insights/instant-payments-in-asia-pacific-ready-for-prime-time

25. Nets. (2014). Nets Launches Real-time 24/7 Payments in Denmark. Retrieved from https://www.nets.eu/en/payments/news/news-2014/Pages/Nets-launches-real-time-24-7-payments-in-Denmark.aspx

26. NPCI. (n.d.). Unified Payments Interface. Retrieved from https://www.npci.org.in/product-overview/upi-product-overview

27. PWC. (n.d.). Instant Payments: Transforming the Gig Economy. Retrieved from https://www.pwc.com/gx/en/industries/financial-services/publications/instant-payments-transforming-gig-economy.html

28. UK Finance. (n.d.). Faster Payments. Retrieved from https://www.ukfinance.org.uk/payments/faster-payments

29. UK Finance. (n.d.). Faster Payments Service. Retrieved from https://www.ukfinance.org.uk/faster-payments-service

30. World Bank. (2020). The Global Shift to Digital Payments. Retrieved from https://blogs.worldbank.org/voices/global-shift-digital-payments

31. Zelle. (n.d.). About Zelle. Retrieved from https://www.zellepay.com/about-zelle

Chapter 6 - Financial Institution Implementation Specifics for Faster Payment Use Cases

1. Atilla Csutak Experience, City National Bank, 2023, https://www.linkedin.com/in/attilacsutak/

Chapter 7 - Risk, Fraud, & Compliance

1. Do Faster Payments Mean Faster Fraud? (securityintelligence.com)

2. Do Faster Payments Mean Faster Fraud? Yes, and No…Part 1 | Bottomline

3. https://www.occ.gov/publications-and-resources/publications/comptrollers-handbook/files/corporate-risk-governance/index-corporate-and-risk-governance.html

4. Ibid

5. https://www.bain.com/insights/how-banks-can-manage-operational-risk/

6. P2P Payment Growth: Just The Tip of The Iceberg? - PaymentsJournal

7. https://www.mckinsey.com/~/media/mckinsey/business%20functions/strategy%20and%20corporate%20finance/our%20insights/how%20covid%2019%20has%20pushed%20companies%20over%20the%20technology%20tipping%20point%20and%20transformed%20business%20forever/how-covid-19-has-pushed-companies-over-the%20technology%20tipping-point-final.pdf

8. US Mobile Peer-to-Peer Payments Forecast 2022 - Insider Intelligence Trends, Forecasts & Statistics

9. Ibid

10. https://files.consumerfinance.gov/f/documents/cfbp_electronic-fund-transfers-faqs.pdf

11. https://www.cnn.com/2022/09/22/economy/bank-ceos-testify-senate/index.html

12. ZELLE REPORT OCTOBER 2022.pdf (senate.gov)

13. https://www.nacha.org/sites/default/files/2022-09/9.22%20Risk%20Management%20Framework.pdf

14. https://www.nacha.org/system/files/2023-05/Executive%20Summary.pdf

15. https://publications.parliament.uk/pa/cm5803/cmselect/cmtreasy/939/summary.html

16. Rules_Interp_Scope_RFP_Warranty_11-30-2022.pdf (theclearinghouse.org)

17. FedNow[SM] Service Operating Procedures (frbservices.org)

18. OFAC-Domestic-ACH-Policy.pdf (nacha.org)

19. FFIEC Issues Guidance on Authentication and Access to Financial Institution Services and Systems | NCUA

20. https://www.occ.gov/publications-and-resources/publications/comptrollers-handbook/files/corporate-risk-governance/index-corporate-and-risk-governance.html

21. Developing a Modern Payments Strategy Begins with Answering These 11 Questions (paymentsjournal.com)

22. https://www.occ.gov/publications-and-resources/publications/comptrollers-handbook/files/corporate-risk-governance/index-corporate-and-risk-governance.html

23. Account Fraud Harder to Detect as Crime Moves from Bots to Sweatshops (darkreading.com)

24. Who's Making All Those Scam Calls? - The New York Times (nytimes.com)

25. https://arstechnica.com/information-technology/2023/01/ microsofts-new-ai-can-simulate-anyones-voice-with-3-seconds-of-audio/

Chapter 8 - Instant Payment rocket boosters: AI, QR codes, Open Banking APIs, Embedded Payments, and Directories

1. US Faster Payments Council QR code White Paper https:// fasterpaymentscouncil.org/userfiles/2080/files/QR%20 Code%20White%20Paper_07-25-2022_Final(2).pdf

2. US Faster Payments Council Directory Models Infographic "Making the case for an Interoperable Directory to facilitate Faster Payments" https://fasterpaymentscouncil. org/userfiles/2080/files/FPC%20DMWG%20 Infographic_01-24-2022%20Final(1).pdf

The underlying whitepapers were:

The Economic Benefits of an Independent Interoperable Directory for Faster payments

https://fasterpaymentscouncil.org/blog/5523/-The-Economic-Benefits-of-an-Independent-Interoperable-Directory-for-Faster-Payments

Beneficial Characteristics Desirable in a Directory Service

https://fasterpaymentscouncil.org/blog/6331/Beneficial-Characteristics-Desirable-in-a-Directory-Service

www.ingramcontent.com/pod-product-compliance
Lightning Source LLC
Chambersburg PA
CBHW061146220326
41599CB00025B/4371